# Biosensors Based on Sandwich Assays

Fan Xia · Xiaojin Zhang
Xiaoding Lou · Quan Yuan
Editors

# Biosensors Based on Sandwich Assays

 Springer

*Editors*
Fan Xia
Engineering Research Center
  of Nano-Geomaterials
  of Ministry of Education, Faculty
  of Materials Science and Chemistry
China University of Geosciences
Wuhan, Hubei
China

Xiaojin Zhang
Engineering Research Center
  of Nano-Geomaterials
  of Ministry of Education, Faculty
  of Materials Science and Chemistry
China University of Geosciences
Wuhan, Hubei
China

Xiaoding Lou
Engineering Research Center
  of Nano-Geomaterials
  of Ministry of Education, Faculty
  of Materials Science and Chemistry
China University of Geosciences
Wuhan, Hubei
China

Quan Yuan
College of Chemistry and Molecular
  Sciences
Wuhan University
Wuhan, Hubei
China

ISBN 978-981-13-4009-3          ISBN 978-981-10-7835-4    (eBook)
https://doi.org/10.1007/978-981-10-7835-4

This Springer imprint is published by Springer Nature
The registered company is Springer Nature Singapore Pte Ltd.
The registered company address is: 152 Beach Road, #21-01/04 Gateway East, Singapore 189721, Singapore

# Foreword

Proteins, nucleic acids, small molecules, ions, and pathogens comprise important components of organic life, and they also serve as indicators of biological growth, development, reproduction, inheritance and variation. Abnormalities of these biomarkers are often reflective of physiological disorder or the presence of disease. In this regard, sandwich assays are key detectors of such biomarkers and, hence, a powerful technique in the fields of clinical diagnostics, molecular detection, and environmental monitoring. It is intriguing that sandwich assays have been extensively developed, right along with advancements in chemistry, biotechnology, and nanotechnology. Accordingly, this book offers a comprehensive understanding of sandwich assays in the context of such advancements, particularly those assays specifically built as biosensors for the many new biomarkers becoming available for disease diagnosis and environmental detection.

Xia, Zhang, Lou, and Yuan, editors of this remarkable book, offer a broad perspective of biosensor development based on sandwich assays, explaining how their applications have expanded since the sandwich assay was first introduced. So far, colorimetric, fluorescence, electrochemical, giant magnetoresistive, localized surface plasmon resonance, surface-enhanced Raman scattering, quartz crystal microbalance, microcantilever, and other novel methods have been employed in the design of the sandwich assays. Some portable devices using microfluidic and electronic technologies based on sandwich assays have already been applied in clinical diagnosis. Some strategies, such as hybridization chain reaction, will further improve the specificity and sensitivity of these assays. This well-organized and well-written treatise will inspire researchers to continue working toward that goal with the prospect of even more significant applications in personalized medicine, clinical diagnosis and other important detection schemes.

Hunan, China

Weihong Tan
Hunan University
University of Florida

*The original version of the book was revised:
Foreword has been included and authors'
affiliations have been updated. The erratum
to this book is available at
https://doi.org/10.1007/978-981-10-7835-4_13*

# Contents

# Editors and Contributors

## About the Editors

**Fan Xia** is currently a Professor at Huazhong University of Science and Technology (HUST) and Dean of the Faculty of Materials Science and Chemistry, China University of Geosciences (Wuhan). He received his B.Sc. degree (2003) from HUST and Ph.D. degree (2008) from the Institute of Chemistry, Chinese Academy of Sciences (ICCAS) (Lei Jiang's group). He then worked as a Postdoctoral Fellow in Prof. Alan J. Heeger's group at the University of California, Santa Barbara. He joined HUST as part of the 1000 Young Talents Program in 2012. His scientific interest focuses on bio-analytical chemistry.

**Xiaojin Zhang** is currently a Professor at China University of Geosciences (Wuhan). He received his B.Sc. (2007) and Ph.D. degrees (2012) from Wuhan University (Renxi Zhuo's group). He then worked as a Postdoctoral Fellow in Prof. Peter X. Ma's group at the University of Michigan, Ann Arbor. He joined China University of Geosciences (Wuhan) in 2016. His scientific interests focus on nanomaterials for theranostics.

**Xiaoding Lou** is currently a Professor at China University of Geosciences (Wuhan). She received her Ph.D. degree (2012) from Wuhan University (Zhen Li's group). She then worked as a Research Associate in Prof. Ben Zhong Tang's group at the Hong Kong University of Science and Technology. She joined China University of Geosciences (Wuhan) in 2017. Her scientific interests focus on the chemical and biosensor field.

**Quan Yuan** is currently a Professor at Wuhan University. She received her B.Sc. degree (2004) from Wuhan University and Ph.D. degree (2009) from Peking University (Chunhua Yan's group). She then worked as a Postdoctoral Fellow in Prof. Weihong Tan's group at the University of Florida. She joined Wuhan University in 2011. Her scientific interests focus on upconversion nanoparticles and DNA nanotechnology.

# Contributors

**Yu Dai** Engineering Research Center of Nano-Geomaterials of Ministry of Education, Faculty of Materials Science and Chemistry, China University of Geosciences, Wuhan, People's Republic of China

**Xiaoxia Hu** College of Chemistry and Molecular Sciences, Wuhan University, Wuhan, People's Republic of China

**Fujian Huang** Engineering Research Center of Nano-Geomaterials of Ministry of Education, Faculty of Materials Science and Chemistry, China University of Geosciences, Wuhan, People's Republic of China

**Hui Li** Engineering Research Center of Nano-Geomaterials of Ministry of Education, Faculty of Materials Science and Chemistry, China University of Geosciences, Wuhan, People's Republic of China

**Shaoguang Li** Engineering Research Center of Nano-Geomaterials of Ministry of Education, Faculty of Materials Science and Chemistry, China University of Geosciences, Wuhan, People's Republic of China

**Meihua Lin** Engineering Research Center of Nano-Geomaterials of Ministry of Education, Faculty of Materials Science and Chemistry, China University of Geosciences, Wuhan, People's Republic of China

**Rui Liu** Engineering Research Center of Nano-Geomaterials of Ministry of Education, Faculty of Materials Science and Chemistry, China University of Geosciences, Wuhan, People's Republic of China

**Xinwen Liu** College of Chemistry and Molecular Sciences, Wuhan University, Wuhan, People's Republic of China

**Xiaoding Lou** Hubei Key Laboratory of Bioinorganic Chemistry & Materia Medica,School of Chemistry and Chemical Engineering, Huazhong Universityof Science and Technology, Wuhan, People's Republic of China; Engineering Research Center of Nano-Geomaterials of Ministry of Education, Faculty of Materials Science and Chemistry, China University of Geosciences, Wuhan, People's Republic of China

**Fan Xia** Hubei Key Laboratory of Bioinorganic Chemistry & Materia Medica, School of Chemistry and Chemical Engineering, Huazhong University of Science and Technology, Wuhan, People's Republic of China; Engineering Research Center of Nano-Geomaterials of Ministry of Education, Faculty of Materials Science and Chemistry, China University of Geosciences, Wuhan, People's Republic of China

**Xiaoqing Yi** Engineering Research Center of Nano-Geomaterials of Ministry of Education, Faculty of Materials Science and Chemistry, China University of Geosciences, Wuhan, People's Republic of China

**Quan Yuan** College of Chemistry and Molecular Sciences, Wuhan University, Wuhan, People's Republic of China

**Shenshan Zhan** Hubei Key Laboratory of Bioinorganic Chemistry & Materia Medica, School of Chemistry and Chemical Engineering, Huazhong University of Science and Technology, Wuhan, People's Republic of China

**Xiaojin Zhang** Engineering Research Center of Nano-Geomaterials of Ministry of Education, Faculty of Materials Science and Chemistry, China University of Geosciences, Wuhan, People's Republic of China

**Pei Zhou** School of Agriculture and Biology, Shanghai Jiao Tong University, Shanghai, People's Republic of China

**Xiaolei Zuo** Institute of Molecular Medicine, Renji Hospital, School of Medicine and School of Chemistry and Chemical Engineering, Shanghai Jiao Tong University, Shanghai, People's Republic of China

# Chapter 1
# Introduction

**Xiaojin Zhang and Fan Xia**

**Abstract** The sandwich assays are one of the mainstays in the fields of clinical diagnostics, molecular detection, and environmental monitoring due to their high specificity and good sensitivity for the detection of analytes. Owing to the development of chemistry and material science, the sandwich assays have been developed vigorously with thousands of published papers to date. To further improve the sensitivity, supersandwich assays emerge as the times require. In this chapter, we will introduce the sandwich assays and briefly discuss the applications of the sandwich assays in the detection of proteins, nucleic acids, small molecules, ions, and cells as well as supersandwich assays. The discussion in detail can be found in subsequent chapters.

**Keywords** Sandwich assays · Detection of analytes · Supersandwich assays
High specificity · Good sensitivity

The original version of this chapter was revised: Foreword has been included and authors' affiliations have been updated. The erratum to this chapter is available at https://doi.org/10.1007/978-981-10-7835-4_13

X. Zhang · F. Xia (✉)
Engineering Research Center of Nano-Geomaterials of Ministry of Education,
Faculty of Materials Science and Chemistry, China University of Geosciences,
Wuhan 430074, People's Republic of China
e-mail: xiafan@cug.edu.cn; xiafan@hust.edu.cn

X. Zhang
e-mail: zhangxj@cug.edu.cn

F. Xia
Hubei Key Laboratory of Bioinorganic Chemistry & Materia Medica, School of Chemistry
and Chemical Engineering, Huazhong University of Science and Technology,
Wuhan 430074, People's Republic of China

© Springer Nature Singapore Pte Ltd. 2018
F. Xia et al. (eds.), *Biosensors Based on Sandwich Assays*,
https://doi.org/10.1007/978-981-10-7835-4_1

1

## 1.1    What Are the Sandwich Assays?

An immunoassay is a biochemical method that detects the presence or concentration of a small molecule or a macromolecule in a solution [1]. The molecule measured by the immunoassay is often called as an "analyte" and is usually a protein, nucleic acid, small molecule, ion, and cell. Since Yalow (a co-winner of the 1997 Nobel Prize in Physiology or Medicine) and Berson developed the first immunoassays in the 1950s, immunoassays have become considerable techniques in the detection of analytes for medical and research purposes [2]. Immunoassays employ a series of different labels such as enzymes, radioactive isotopes, DNA reporters, fluorogenic reporters, electrochemiluminescent tags. In addition, there are some kinds of label-free immunoassays which do not require labeling the components of the assay such as surface plasmon resonance technique. Immunoassays can be classified into different formats including (A) competitive immunoassays, (B) one-site, noncompetitive immunoassays, (C) two-site, noncompetitive immunoassays, as shown in Fig. 1.1. In a competitive immunoassay, unlabeled analyte competes with the labeled analyte. The unbound, labeled analyte or the bound, remaining labeled analyte is then measured. In a one-site, noncompetitive immunoassay, the analyte binds with recognition molecule labeled with a detectable signal. The intensity of the bound signal is relative to the amount of analyte. In a two-site, noncompetitive immunoassay, the analyte is bound to recognition molecule 1, then recognition molecule 2 labeled with a detectable signal is bound to the analyte. This type is also referred to as sandwich assay as the binding is a sandwich complex. The sandwich assays usually provide high specificity and good sensitivity for the detection of analytes; therefore, they have become a powerful technique in the fields of clinical diagnostics, molecular detection, and environmental monitoring [3].

## 1.2    Categories of the Sandwich Assays

The sandwich assays can be classified into a number of categories, such as sandwich radioimmunoassay (RIA), sandwich colorimetric assay (CMA), sandwich fluorescence immunoassay (FLIA), sandwich electrochemical assay (ELCA), sandwich giant magnetoresistive assay (GMRA), and sandwich localized surface plasmon resonance assay (LSPRA), as shown in Fig. 1.2 [4]. These assays have the advantages and limitations. For instance, sandwich radioimmunoassay is extremely specific and extremely sensitive, but requires specialized instruments [5]. Sandwich colorimetric assay affords a low detection limit, but needs a long incubation time [6]. Sandwich fluorescence immunoassay has high detection sensitivity, but the stability of the dyes is often not good [7]. Since the electrochemical signaling has high sensitivity, sandwich electrochemical assay obtains the lowest detection limit in comparison with other sandwich assays [8]. Sandwich giant magnetoresistive assay is the most rapid detection method with a short analysis time [9]. Sandwich

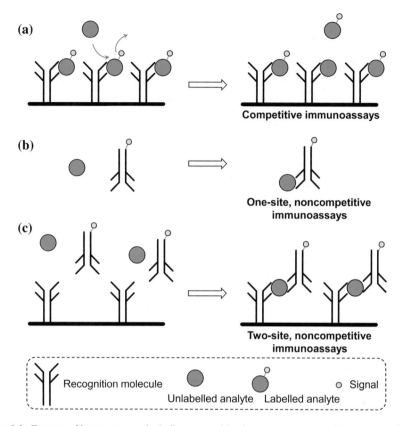

**(a)**

Competitive immunoassays

**(b)**

One-site, noncompetitive immunoassays

**(c)**

Two-site, noncompetitive immunoassays

Recognition molecule    Unlabelled analyte    Labelled analyte    ○ Signal

**Fig. 1.1** Formats of immunoassays including competitive immunoassays, one-site, noncompetitive immunoassays, and two-site, noncompetitive immunoassays also referred to as sandwich assays

localized surface plasmon resonance assay allows label-free detection of analytes, but low concentration analytes could not be detected directly [10]. Along with the development of chemistry, biotechnology, and nanotechnology, numerous efforts that successfully develop the sandwich assays have been reported to date for the detection of analytes, including proteins, nucleic acids, small molecules, ions, and cells [11]. Furthermore, supersandwich assays that integrate multiple signal probes together have also been developed to amplify the signal and lower the detection limit [12]. In the following part, we will discuss the applications of the sandwich assays in the detection of proteins, nucleic acids, small molecules, ions, and cells as well as supersandwich assays.

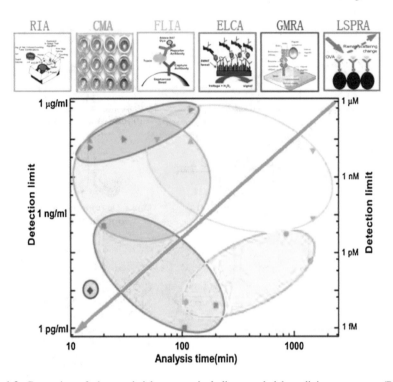

**Fig. 1.2** Categories of the sandwich assays including sandwich radioimmunoassay (RIA), sandwich colorimetric assay (CMA), sandwich fluorescence immunoassay (FLIA), sandwich electrochemical assay (ELCA), sandwich giant magnetoresistive assay (GMRA), and sandwich localized surface plasmon resonance assay (LSPRA) (Reprinted with the permission from Ref. [4]. Copyright 2014 American Chemical Society)

## 1.3  Sandwich Assays for Protein Detection

Proteins, one important class of biomarkers, are indicative of normal or disease-related biological processes as well as the onset, existence, or progression of cancer; e.g., serum prostate-specific antigen (PSA) is overexpressed in some prostate diseases as well as prostate cancer [13]. The development of powerful, cost-effective, reliable detection and monitoring techniques for proteins is particularly important [14]. Up to now, most clinical protein detection is performed through an enzyme-linked immunosorbent assay (ELISA), but relatively expensive kits and equipment are required [15]. The sandwich assays using colorimetric [16], fluorescence [17], electrochemical [18], electrochemiluminescence [19], giant magnetoresistive [20], and localized surface plasmon resonance [21] detection methods are employed to reduce detection cost and simplify operation. For example, Ju et al. designed a ratiometric electrochemical assay via immunoreaction-induced DNA assembly on the surface of Au electrode for the detection of protein, as shown in Fig. 1.3 [22]. The target protein triggered the sandwich

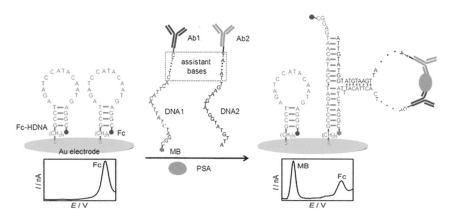

**Fig. 1.3** Schematic illustration of ratiometric electrochemical immunoassay for the detection of protein. A ferrocene (Fc)-labeled hairpin DNA was immobilized on the surface of gold electrode. The target protein (PSA) induced the sandwich immunoreaction to trigger the hybridization of DNA1 and DNA2 with antibody1 and antibody2, respectively (Reprinted with the permission from Ref. [22]. Copyright 2014 Elsevier)

immunoreaction of two antibody-modified DNA probes. The result caused Fc departure from the electrode and MB approach to the electrode. Two signals of MB and Fc were read for the detection of PSA with a detection limit of 4.3 pg mL$^{-1}$.

## 1.4  Sandwich Assays for Nucleic Acid Detection

Nucleic acids discovered by Friedrich Miescher in 1869 including the DNA and the RNA are essential and the most important biological molecules for all known forms of life [23]. The functions of nucleic acids contain genetic information storing, genetic message transferring, protein synthesis, etc. [24]. The importance of nucleic acids within living cells is undisputed. The changes in the DNA sequence cause most of the genetic disorders that affect humans and other organisms [25]. The detection and analysis of nucleic acids is a significant and rapidly growing field in biomedical research for the application in disease diagnosis, epigenetic, human identification, etc. [26]. Some strategies such as isothermal amplification technology [27], field effect [28], chemistry-driven approach [29], microbead-based technology [30], nanopore-based technique [31], nanotechnology [32], and fluorescent hybridization probe [33] have been adopted for sensitive and specific detection of nucleic acids. Taking an example of the sandwich assays, Zhang et al. reported a lateral flow nucleic acid biosensor for visual detection of miRNA, as shown in Fig. 1.4 [34]. Gold nanoparticle (AuNP)-labeled DNA probe, miRNA, and biotin-modified DNA probe are hybridized on the lateral flow device to form supersandwich DNA structure. The visual red band on the test line is derived from

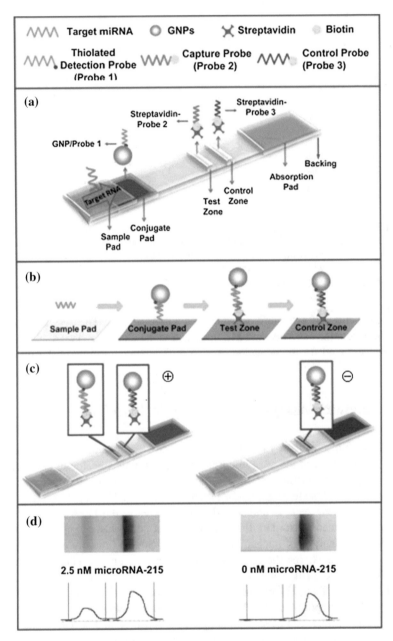

**Fig. 1.4 a** Schematic illustration of the configuration and measurement principle of the lateral flow nucleic acid biosensor; **b** schematic illustration of the sandwich assay procedure; **c** the principle of visual detection in the presence (left) and absence (right) of miRNA-215; **d** typical photographic images and recorded optical responses of lateral flow nucleic acid biosensor in the presence (2.5 nM) and absence (0 nM) of miRNA-215 (Reprinted with the permission from Ref. [34]. Copyright 2013 Elsevier)

the accumulation of AuNPs. The intensity of the bands quantitated by a portable strip reader was relative to the concentration of target miRNA. The assay was able to detect miRNA-215 from A549 cells with a detection limit of 60 pM.

## 1.5 Sandwich Assays for Small Molecule and Ion Detection

Small molecules with molecular weight usually less than 1 kDa are the major analytes of biomedical, food, and environmental fields [35]. Metal ions play a critical role in biological and environmental systems, but cause various health problems if they accumulate in human body such as lead ($Pb^{2+}$), copper ($Cu^{2+}$), mercury ($Hg^{2+}$), and silver ($Ag^+$) ions [36]. Small organic molecules (such as ATP, glucose, cysteine, homocysteine, trinitrotoluene, cocaine, melamine, dopamine) and ions including cations (such as $Hg^{2+}$, $Cu^{2+}$, $Ca^{2+}$, $Pb^{2+}$, $As^{3+}$, $Al^{3+}$) and anions (such as $NO_2^-$, $F^-$, $I^-$, $CN^-$, $PF_6^-$, oxoanions) have been detected through the colorimetric and fluorescent assays [37] as well as surface-enhanced Raman scattering methods [38], photoelectrochemical assays [39]. For example, Xiao et al. reported nanoprobe-enhanced, split aptamer-based electrochemical sandwich assay for ultrasensitive detection of small molecules such as ATP and cocaine, as shown in Fig. 1.5 [40]. In the conventional split aptamer-based electrochemical sandwich assay, the reporter probes either create target-capture probe complex or adsorb on the surface of the electrode, leading to high background and low sensitivity. In the nanoprobe-enhanced, split aptamer-based electrochemical sandwich assay, nanoreporter probes immobilized on AuNPs achieve enhanced binding affinity and reduced probe concentration. The multiple bindings generate near-zero background and improved sensitivity.

## 1.6 Sandwich Assays for Cell Detection

Early detection of cells, particularly cancer cells, is of importance to enhanced prognosis and cancer management due to small tumor size and unknown site of the tumor [41]. Circulating tumor cells (CTCs) are shed from primary tumors into blood circulation and are in extremely sparse content as few as 1 cell in 1 mL of patient blood [42]. CTCs as seeds may lead to the subsequent growth of additional tumors, and thus, the capture and detection of CTCs have the clinical significance in modern cancer research [43]. Various nanotechnology-based strategies to achieve high sensitivity and low detection limit for the detection of CTCs have been significantly developed in the past years [44–46]. For instance, Xu et al. proposed an electrochemiluminescence biosensor with a closed bipolar electrode for the detection of cancer cells, as shown in Fig. 1.6 [47]. Two-channel polydimethylsiloxane

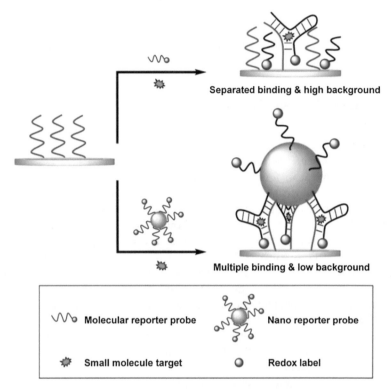

**Fig. 1.5** In conventional split aptamer-based electrochemical sandwich assay (top), target-capture probe complexes or nonspecifically adsorb onto the electrode surface leads to high background. Nanoreporter probe comprising gold nanoparticles (AuNPs) conjugated with multiple molecular reporter probes yields elevated binding affinity and low background (Reprinted with the permission from Ref. [40]. Copyright 2015 American Chemical Society)

(PDMS) chip was connected using U-shaped indium tin oxide (ITO) on the surface of a glass. At the cathode, folic acid recognized cancer cells with folate receptor at cancer cell membrane. Graphene–Au conjugates with aptamers increased the electroconductivity of the cathode. The catalytic reduction of oxygen at the cathode enhanced the ECL signal. Therefore, this strategy achieved a detection limit of as low as 18 cells in 30 μL of cell suspension.

## 1.7 Supersandwich Assays

Since a target only integrates with one signal probe in the traditional sandwich assays, the sensitivity is restricted. To overcome this limitation, supersandwich assays that bind multiple signal probes together to improve the total signal have been developed in the past decades [48–53]. The classic case of supersandwich

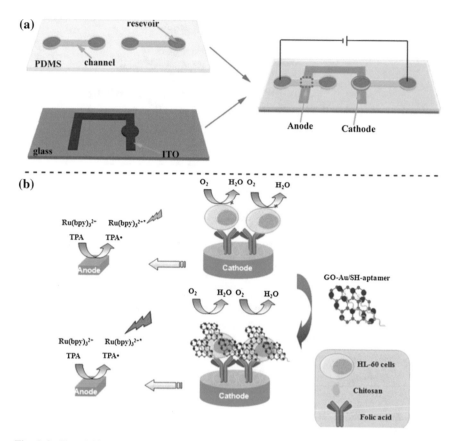

**Fig. 1.6** Closed bipolar electrode–electrochemiluminescence assay for the detection of cancer cells. **a** Schematic illustration of the biosensing principle; **b** a sandwich-type assay at the cathodic pole was constructed by using folic acid to capture cancer cells and GO-Au.SH-aptamer to increase the electroconductivity of the cathode (Reprinted with the permission from Ref. [47]. Copyright 2016 Wiley-VCH Verlag GmbH & Co. KGaA, Weinheim)

assays was developed by Xia, Zuo, Plaxco, and Heeger in 2010 through DNA hybridization to create long concatamers with multiple target molecules and signal probes for amplifying the signal and lowering the detection limit [48]. Xia et al. then constructed supersandwich DNA structure in the nanopores for the detection of $Zn^{2+}$ with a detection limit of as low as 1 nM, as shown in Fig. 1.7 [54]. The capture probe anchored onto the nanopores captured the sessile probe followed by hybridization of two auxiliary probes to create supersandwich DNA structure in the nanopores. DNAzyme strand is then hybridized with the sessile probe to generate a DNAzyme system. With the addition of $Zn^{2+}$, DNAzyme strand cut the sessile probe, leading to the dissociation of supersandwich DNA structure. The ion current through the nanopores increased relative to the concentration of $Zn^{2+}$.

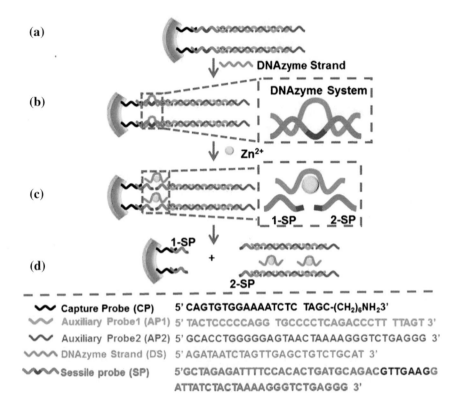

**Fig. 1.7** Supersandwich assay for the detection of $Zn^{2+}$. **a** The formation of the DNA supersandwich structures on the internal surface of a nanopore by the bridging of the CPs and SPs of the substrate; **b** the formation of the DNAzyme system through partial hybridization between the $Zn^{2+}$-requiring DNAzyme and the SPs; **c** the fragmentation of the SPs into two parts in the presence of $Zn^{2+}$; and **d** the peeling of the DNA supersandwich structures from the internal surface of the nanopores (Reprinted with the permission from Ref. [54]. Copyright 2015 The Royal Society of Chemistry)

## 1.8  Conclusion

In this chapter, we provided the introduction and the applications of the sandwich assays in the detection of proteins, nucleic acids, small molecules, ions, and cells as well as supersandwich assays. Read subsequent chapters to obtain more details. It is clear that the sandwich assays reported in recent years have reached a very high level with an extremely low detection limit as low as signal molecule. In fact, the practical application such as portable diagnostic devices based on the sandwich assays is more urgent, particularly in some developing countries or remote areas [55]. It is encouraging that lateral flow tests have been successfully commercialized for the detection of proteins [56–58]. Some other portable devices using microfluidic and electronic technologies have been developed too [59–61]. It should be noted that reagents are often necessary in the detection process and the stability of the sandwich complex is still a problem.

# References

1. Zhao LX, Sun L, Chu XG (2009) Chemiluminescence immunoassay. TrAC-trends. Anal Chem 28:404–415
2. Fu XL, Chen LX, Choo J (2017) Optical nanoprobes for ultrasensitive immunoassay. Anal Chem 89:124–137
3. Pei XM, Zhang B, Tang J, Liu BQ, Lai WQ, Tang DP (2013) Sandwich-type immunosensors and immunoassays exploiting nanostructure labels: a review. Anal Chim Acta 758:1–18
4. Shen JW, Li YB, Gu HS, Xia F, Zuo XL (2014) Recent development of sandwich assay based on the nanobiotechnologies for proteins, nucleic acids, small molecules, and ions. Chem Rev 114:7631–7677
5. Liu R, Zhang SX, Wei C, Xing Z, Zhang SC, Zhang XR (2016) Metal stable isotope tagging: renaissance of radioimmunoassay for multiplex and absolute quantification of biomolecules. Acc Chem Res 49:775–783
6. Yin YM, Cao Y, Xu YY, Li GX (2010) Colorimetric immunoassay for detection of tumor markers. Int J Mol Sci 11:5078–5095
7. Smith DS, Eremin SA (2008) Fluorescence polarization immunoassays and related methods for simple, high-throughput screening of small molecules. Anal Bioanal Chem 391:1499–1507
8. Fojta M, Danhel A, Havran L, Vyskocil V (2016) Recent progress in electrochemical sensors and assays for DNA damage and repair. TrAC-trends Anal Chem 79:160–167
9. Wang SX, Li G (2008) Advances in giant magnetoresistance biosensors with magnetic nanoparticle tags: review and outlook. IEEE Trans Magn 44:1687–1702
10. Unser S, Bruzas I, He J, Sagle L (2015) Localized surface plasmon resonance biosensing: current challenges and approaches. Sensors 15:15684–15716
11. Teste B, Descroix S (2012) Colloidal nanomaterial-based immunoassay. Nanomedicine 7:917–929
12. Liu NN, Huang FJ, Lou XD, Xia F (2017) DNA hybridization chain reaction and DNA supersandwich self-assembly for ultrasensitive detection. Sci China-Chem 60:311–318
13. Rusling JF, Kumar CV, Gutkind JS, Patel V (2010) Measurement of biomarker proteins for point-of-care early detection and monitoring of cancer. Analyst 135:2496–2511
14. Zhang Y, Guo YM, Xianyu YL, Chen WW, Zhao YY, Jiang XY (2013) Nanomaterials for ultrasensitive protein detection. Adv Mater 25:3802–3819
15. Chikkaveeraiah BV, Bhirde AA, Morgan NY, Eden HS, Chen XY (2012) Electrochemical immunosensors for detection of cancer protein biomarkers. ACS Nano 6:6546–6561
16. Chen CH, Luo M, Ye T, Li NX, Ji XH, He ZK (2015) Sensitive colorimetric detection of protein by gold nanoparticles and rolling circle amplification. Analyst 140:4515–4520
17. Wang B, Yu C (2010) Fluorescence turn-on detection of a protein through the reduced aggregation of a perylene probe. Angew Chem Int Ed 49:1485–1488
18. Lai GS, Yan F, Ju HX (2009) Dual signal amplification of glucose oxidase-functionalized nanocomposites as a trace label for ultrasensitive simultaneous multiplexed electrochemical detection of tumor markers. Anal Chem 81:9730–9736
19. Zhu DB, Hou XM, Xing D (2012) Ultrasensitive aptamer-based bio bar code immunomagnetic separation and electrochemiluminescence method for the detection of protein. Anal Chim Acta 725:39–43
20. Wang T, Yang Z, Lei C, Lei J, Zhou Y (2014) An integrated giant magnetoimpedance biosensor for detection of biomarker. Biosens Bioelectron 58:338–344

21. Wu B, Jiang R, Wang Q, Huang J, Yang XH, Wang KM, Li WS, Chen ND, Li Q (2016) Detection of C-reactive protein using nanoparticle-enhanced surface plasmon resonance using an aptamer-antibody sandwich assay. Chem Commun 52:3568–3571

22. Ren KW, Wu J, Yan F, Zhang Y, Ju HX (2015) Immunoreaction-triggered DNA assembly for one-step sensitive ratiometric electrochemical biosensing of protein biomarker. Biosens Bioelectron 66:345–349

23. Dahm R (2008) Discovering DNA: Friedrich Miescher and the early years of nucleic acid research. Hum Genet 122:565–581

24. Nakano S, Miyoshi D, Sugimoto N (2014) Effects of molecular crowding on the structures, interactions, and functions of nucleic acids. Chem Rev 114:2733–2758

25. Wachowius F, Attwater J, Holliger P (2017) Nucleic acids: function and potential for abiogenesis. Q Rev Biophys 50:1–37

26. Gerasimova YV, Kolpashchikov DM (2014) Enzyme-assisted target recycling (EATR) for nucleic acid detection. Chem Soc Rev 43:6405–6438

27. Safavieh M, Kanakasabapathy MK, Tarlan F, Ahmed MU, Zourob M, Asghar W, Shafiee H (2016) Emerging loop-mediated isothermal amplification-based microchip and microdevice technologies for nucleic acid detection. ACS Biomater Sci Eng 2:278–294

28. Veigas B, Fortunato E, Baptista PV (2015) Field effect sensors for nucleic acid detection: recent advances and future perspectives. Sensors 15:10380–10398

29. Smith SJ, Nemr CR, Kelley SO (2017) Chemistry-driven approaches for ultrasensitive nucleic acid detection. J Am Chem Soc 139:1020–1028

30. Rodiger S, Liebsch C, Schmidt C, Lehmann W, Resch-Genger U, Schedler U, Schierack P (2014) Nucleic acid detection based on the use of microbeads: a review. Microchim Acta 181:1151–1168

31. Ying YL, Zhang JJ, Gao R, Long YT (2013) Nanopore-based sequencing and detection of nucleic acids. Angew Chem Int Ed 52:13154–13161

32. Hartman MR, Ruiz RCH, Hamada S, Xu CY, Yancey KG, Yu Y, Han W, Luo D (2013) Point-of-care nucleic acid detection using nanotechnology. Nanoscale 5:10141–10154

33. Guo J, Ju JY, Turro NJ (2012) Fluorescent hybridization probes for nucleic acid detection. Anal Bioanal Chem 402:3115–3125

34. Gao XF, Xu H, Baloda M, Gurung AS, Xu LP, Wang T, Zhang XJ, Liu GD (2014) Visual detection of microRNA with lateral flow nucleic acid biosensor. Biosens Bioelectron 54:578–584

35. Shankaran DR, Gobi KVA, Miura N (2007) Recent advancements in surface plasmon resonance immunosensors for detection of small molecules of biomedical, food and environmental interest. Sens Actuator B-Chem 121:158–177

36. Huang JH, Su XF, Li ZG (2017) Metal ion detection using functional nucleic acids and nanomaterials. Biosens Bioelectron 96:127–139

37. Liu DB, Wang Z, Jiang XY (2011) Gold nanoparticles for the colorimetric and fluorescent detection of ions and small organic molecules. Nanoscale 3:1421–1433

38. Alvarez-Puebla RA, Liz-Marzan LM (2012) SERS detection of small inorganic molecules and ions. Angew Chem Int Ed 51:11214–11223

39. Zhao WW, Xu JJ, Chen HY (2016) Photoelectrochemical detection of metal ions. Analyst 141:4262–4271

40. Zhao T, Liu R, Ding XF, Zhao JC, Yu HX, Wang L, Xu Q, Wang X, Lou XH, He M, Xiao Y (2015) Nanoprobe-enhanced, split aptamer-based electrochemical sandwich assay for ultrasensitive detection of small molecules. Anal Chem 87:7712–7719

41. Chen J, Li J, Sun Y (2012) Microfluidic approaches for cancer cell detection, characterization, and separation. Lab Chip 12:1753–1767

42. Arya SK, Lim B, Rahman ARA (2013) Enrichment, detection and clinical significance of circulating tumor cells. Lab Chip 13:1995–2027

43. Castro CM, Ghazani AA, Chung J, Shao HL, Issadore D, Yoon TJ, Weissleder R, Lee H (2014) Miniaturized nuclear magnetic resonance platform for detection and profiling of circulating tumor cells. Lab Chip 14:14–23

44. Yu L, Ng SR, Xu Y, Dong H, Wang YJ, Li CM (2013) Advances of lab-on-a-chip in isolation, detection and post-processing of circulating tumour cells. Lab Chip 13:3163–3182
45. Alix-Panabieres C, Pantel K (2014) Technologies for detection of circulating tumor cells: facts and vision. Lab Chip 14:57–62
46. Lin M, Chen JF, Lu YT, Zhang Y, Song JZ, Hou S, Ke ZF, Tseng HR (2014) Nanostructure embedded microchips for detection, isolation, and characterization of circulating tumor cells. Acc Chem Res 47:2941–2950
47. Wu MS, Liu Z, Xu JJ, Chen HY (2016) Highly specific electrochemiluminescence detection of cancer cells with a closed bipolar electrode. ChemElectroChem 3:429–435
48. Xia F, White RJ, Zuo XL, Patterson A, Xiao Y, Kang D, Gong X, Plaxco KW, Heeger AJ (2010) An electrochemical supersandwich assay for sensitive and selective DNA detection in complex matrices. J Am Chem Soc 132:14346–14348
49. Liu NN, Jiang YN, Zhou YH, Xia F, Guo W, Jiang L (2013) Two-way nanopore sensing of sequence-specific oligonucleotides and small-molecule targets in complex matrices using integrated DNA supersandwich structures. Angew Chem Int Ed 52:2007–2011
50. Wei BM, Liu NN, Zhang JT, Ou XW, Duan RX, Yang ZK, Lou XD, Xia F (2015) Regulation of DNA self-assembly and DNA hybridization by chiral molecules with corresponding biosensor applications. Anal Chem 87:2058–2062
51. Wei BM, Zhang JT, Wang HB, Xia F (2016) A new electrochemical aptasensor based on a dual-signaling strategy and supersandwich assay. Analyst 141:4313–4318
52. Wei BM, Zhang TC, Ou XW, Li XC, Lou XD, Xia F (2016) Stereochemistry-guided DNA probe for single nucleotide polymorphisms analysis. ACS Appl Mater Interfaces 8:15911–15916
53. Jiang YN, Liu NN, Guo W, Xia F, Jiang L (2012) Highly-efficient gating of solid-state nanochannels by DNA supersandwich structure containing ATP aptamers: a nanofluidic IMPLICATION logic device. J Am Chem Soc 134:15395–15401
54. Liu NN, Hou RZ, Gao PC, Lou XD, Xia F (2016) Sensitive $Zn^{2+}$ sensor based on biofunctionalized nanopores via combination of DNAzyme and DNA supersandwich structures. Analyst 141:3626–3629
55. Khanna P, Walt DR (2015) Salivary diagnostics using a portable point-of-service platform: a review. Clin Ther 37:498–504
56. Fenton EM, Mascarenas MR, Lopez GP, Sibbett SS (2009) Multiplex lateral-flow test strips fabricated by two-dimensional shaping. ACS Appl Mater Interfaces 1:124–129
57. Roder M, Vieths S, Holzhauser T (2009) Commercial lateral flow devices for rapid detection of peanut (Arachis hypogaea) and hazelnut (Corylus avellana) cross-contamination in the industrial production of cookies. Anal Bioanal Chem 395:103–109
58. Bamrungsap S, Apiwat C, Chantima W, Dharakul T, Wiriyachaiporn N (2014) Rapid and sensitive lateral flow immunoassay for influenza antigen using fluorescently-doped silica nanoparticles. Microchim Acta 181:223–230
59. Barbosa AI, Gehlot P, Sidapra K, Edwards AD, Reis NM (2015) Portable smartphone quantitation of prostate specific antigen (PSA) in a fluoropolymer microfluidic device. Biosens Bioelectron 70:5–14
60. Chen M, Yang H, Rong LY, Chen XQ (2016) A gas-diffusion microfluidic paper-based analytical device (mu PAD) coupled with portable surface-enhanced Raman scattering (SERS): facile determination of sulphite in wines. Analyst 141:5511–5519
61. Liu D, Li XR, Zhou JK, Liu SB, Tian T, Song YL, Zhu Z, Zhou LJ, Ji TH, Yang CY (2017) A fully integrated distance readout ELISA-Chip for point-of-care testing with sample-in-answer-out capability. Biosens Bioelectron 96:332–338

# Chapter 2
# Colorimetric Sandwich Assays for Protein Detection

Xiaoqing Yi, Rui Liu, Xiaoding Lou and Fan Xia

**Abstract** For decades, the sandwich assays have been widely used in quality control, clinical diagnostics, biological detection, and environmental monitoring field. Apart from the requirement of labeling the molecular target, the sandwich assay generally requires the recognition and signaling probe to be combined, which gives them accurate specific. Also, the optical sensing method is of great interest due to the intrinsic high sensitivity and simplicity. The colorimetric assays are much simpler for the detection of analytes, and their response can be directly detected with bare eyes or by photometry compared with other analytical techniques. Colorimetric sandwich assays are usually based on the observable color variation in the presence of enzyme-labeled antibody. In this chapter, we focus on the detection of protein by colorimetric sandwich assays based on traditional enzymes and bio-mimetic nanomaterials. The detection technologies employed in colorimetric sandwich assays based on traditional enzymes were alkaline phosphatase (ALP),

The original version of this chapter was revised: Foreword has been included and authors' affiliations have been updated. The erratum to this chapter is available at https://doi.org/10.1007/978-981-10-7835-4_13

X. Yi · R. Liu · X. Lou (✉) · F. Xia
Engineering Research Center of Nano-Geomaterials of Ministry of Education,
Faculty of Materials Science and Chemistry, China University of Geosciences,
Wuhan 430074, People's Republic of China
e-mail: louxiaoding@cug.edu.cn; louxiaoding@hust.edu.cn

X. Yi
e-mail: keyi0115@126.com

R. Liu
e-mail: liurui2016@sohu.com

F. Xia
e-mail: xiafan@cug.edu.cn; xiafan@hust.edu.cn

X. Lou · F. Xia
Hubei Key Laboratory of Bioinorganic Chemistry & Materia Medica,
School of Chemistry and Chemical Engineering, Huazhong University of Science
and Technology, Wuhan 430074, People's Republic of China

© Springer Nature Singapore Pte Ltd. 2018
F. Xia et al. (eds.), *Biosensors Based on Sandwich Assays*,
https://doi.org/10.1007/978-981-10-7835-4_2

15

horseradish peroxidase (HRP), open sandwich immunoassay (OS-IA), and gold–multienzyme–nanocarrier. For the biomimetic nanomaterials, highlighted examples were focused on $Fe_3O_4$ magnetic nanoparticles (MNPs) biomimetic enzymes, and Au@Pt nanostructure biomimetic enzymes.

**Keywords** Colorimetric sandwich assays · Protein detection · Biomimetic enzymes · Traditional enzymes · Nanoparticles

## 2.1 Introduction

Various techniques have been used for the detection of protein, including capillary electrophoresis [1], high-performance liquid chromatography (HPLC) [2], and mass spectrometry (MS) [3]. Generally, these analytical methods are expensive and time-consuming and require specialized operators. In 1971, Engvall and Perlmann developed a user-friendly and enzyme-driven colorimetric assay, now known as the enzyme-linked immunosorbent assay (ELISA) [4]. The platform of the sandwich colorimetric assay is shown in Fig. 2.1. The target is first combined with the surface-bound capture antibody, and then, the added second enzyme-labeled antibody forms the sandwich-like immunocomplex. After introducing the substrate of the labeled enzyme, the variation of the test sample color can be used for quantitative target antigen. The sandwich format colorimetric assay could enhance not only the sensitivity but also the specificity [5].

Optical sensing methods are of great interest due to the intrinsic high sensitivity and simplicity. Compared with other analytical techniques, the colorimetric assays are much simpler for the detection of analytes and their response can be directly

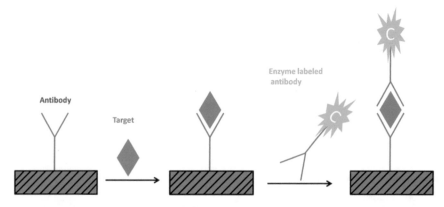

**Fig. 2.1** Principle of colorimetric sandwich assays. The target is first combined with the surface-bound capture antibody, and then, the added second enzyme-labeled antibody forms the sandwich-like immunocomplex. After introducing the substrate of the labeled enzyme, the variation of the test sample color can be used for quantitative target antigen

detected with bare eyes or by photometry. Colorimetric sandwich assays are usually based on the observable color changes in the presence of enzyme-labeled antibody. The colorimetric sandwich assays utilize a monoclonal antibody directed against a distinct antigenic determinant on the intact protein molecule and used for solid-phase immobilization, and an enzyme with a secondary antibody is used as a signal generator. The target proteins are allowed to react sequentially with the two antibodies in the test sample, leading to sandwiching of the antigen between the solid-phase and enzyme-linked antibodies. These linked enzymes can react with a variety of substrates in the reaction to produce measurable spectral signals that can be used to measure the amount of target proteins in the sample (Fig. 2.1). Similarly, colorimetric sandwich assays based on traditional enzymes have been widely used in medical and plant pathology diagnostics, food and environmental residue analysis, and quality control in various industries [6, 7]. Peroxidase, e.g., horseradish peroxidase (HRP), is a protein-based enzyme that acts as a catalyst to promote various biological processes, and is usually adopted in the colorimetric sandwich assays [8]. Moreover, owing to the rapid development of nanomaterial disciplines, the sensitivity of colorimetric sandwich assays based on nanomaterials has greatly improved. For example, Au nanoparticles (AuNPs) have the advantages of narrow size distribution, good biocompatibility, easy to modify functional groups, and other advantages, which are currently the most practical multienzyme carriers for the immobilization of multienzymes [9–11].

However, the biological activity of most natural enzymes is susceptible to external conditions such as temperature and pH [12]. Nanomaterial-based enzymes have attracted great interest because they have many advantages over natural enzymes, such as stable in a wide range of pH, temperature, and inertia of several kinds of protease [13–17]. Several kinds of nanoparticles have been found to show novel performance with horseradish peroxidase (HRP) and exhibit great potential in colorimetric sandwich assays in recent years. According to the fact that colorimetric sandwich assays are typically based on the observable color change in the presence of enzyme-labeled antibody or nanomaterial-based enzymes, we divide the assays into two categories: based on traditional enzymes and nanomaterial-based enzymes. In each category, the highlighted examples are classified on the basis of the different types of enzyme-labeled antibody or nanomaterial-based enzymes.

## 2.2  Colorimetric Sandwich Assays Based on the Traditional Enzymes

### 2.2.1  Alkaline Phosphatase (ALP) for Colorimetric Sandwich Assays

For the colorimetric sandwich assay, alkaline phosphatase (ALP) and horseradish peroxidase (HRP) are usually used as enzymes to build the sandwich structure

**Table 2.1** Commonly used enzymes for protein detection

| Enzyme | Substrate | Protein | Signal | Detection limit |
|--------|-----------|---------|--------|-----------------|
| ALP | pNPP | SR protein kinase 1 [20] | $\lambda = 405$ nm | 0.55 mg/mL |
| | | Human interleukin-8 [19] | $\lambda = 405$ nm | 16 pg/mL |
| HPR | TMB | $\tau$-Protein [21–25] | $\lambda = 450$ nm | 5 pg/mL |
| | | Insulin-like growth factor [26] | $\lambda = 450$ nm | 0.32 ng/mL |
| | | Coconut milk proteins [22] | $\lambda = 450$ nm | 0.39 ng/mL |
| | | Carcinoembryonic protein [27] | $\lambda = 370$ nm | 0.02 ng/mL |
| | | Thrombin [28] | $\lambda = 450$ nm | 25 fM |
| | oPD | $\alpha$-Fetoprotein [27] | $\lambda = 490$ nm | 0.02 ng/mL |
| | ABTS | Chymotrypsin [29] | $\lambda = 405$ nm | 127 ng/mL |

(Table 2.1). ALP is a hydrolase enzyme aiming to remove phosphate groups from several types of phosphomonoesters, such as p-aminophenyl phosphate, phenyl phosphate, 1-naphthyl phosphate, 2-phospho-L-ascorbic acid, and p-nitrophenyl phosphate (pNPP). As the name implies, alkaline phosphatase is most effective in the alkaline environment of pH 8–10 [18]. Dephosphorylation often results in a change in the reagent color, which in turn provides light signals indicating the presence and concentration of target proteins. For example, ALP could remove the phosphate group from pNPP, which is a common alkaline phosphatase substrate, and the yellow water-soluble product formation from dephosphorylated products with a maximum absorption peak at around 405 nm [19].

### 2.2.1.1 Horseradish Peroxidase (HRP) for Colorimetric Sandwich Assays

HRP is an oxidase responsible for catalyzing the oxidation of a range of organic and inorganic substrates with $H_2O_2$, such as o-phenylenediamine (oPD), 3,3′,5,5′-tetramethylbenzidine (TMB), and 2,2′-azinobis(3-ethylbenzthiazoline-6-sulfonic acid) (ABTS). HRP is the most commonly used one due to its smallest molecular structure, highest stability, and fastest catalytic rate. Hah et al. developed a colorimetric sandwich-type assay based on enzyme-linked aptamer assay to detect as low as 25 fM of thrombin with high linearity fast and sensitively [28]. Aptamer-immobilized glass was employed to capture the target analyte, while a second aptamer, functionalized with HRP, was utilized for the conventional TMB-based colorimetric detection. Harris et al. developed a HRP-based colorimetric sandwich assays for the detection of Shiga toxins 1 and 2 (i.e., Stx1 and Stx2) from Shiga toxin-producing E. coli (STEC) bacteria [30]. However, the way with the 1:1 enzyme-antibody conjugate is inherently limited to further improve the detection limit. He et al. first developed an amplified colorimetric method to detect proteins and cancer cells based on the assembly of nucleic acids and proteins, as shown in Fig. 2.2 [31]. The biotinylated DNA strand was connected with sandwich

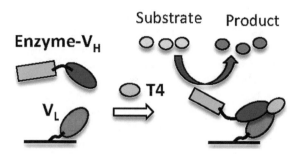

**Fig. 2.2** Schematic of the ultrasensitive colorimetric protein assay based on the assembly of nucleic acids and proteins. The capture antibody was first immobilized on the 96-well microplate via physical adsorption and blocked with BSA. When the target protein exists and was recognized by the biotin-antibody, then streptavidin (SA) was conjugated to the biotin–antibody through the biotin–SA reaction. The linker DNA hybridized with the bound Bio-H1 to open the hairpin of Bio-H1, and the newly exposed sticky end of Bio-H1 opened the hairpin of Bio-H2 to expose a sticky end on Bio-H2. This sticky end was identical in sequence to the initiator strands and hybridized with the hairpin of Bio-H1. In this way, the assembly of nucleic acid (HCR) process began, which made the self-assembled products of nucleic acids connect to the immunocomplex. After that, the assembly of SA-HRP and Biotin-BSA occurred at the node of nucleic acid assembled products where the biotin moieties were located. During the formation of the complex, Biotin-BSA served as a bridge between SA-HRP to form an aggregate of SA-HRP. Under the catalysis of HRP, TMB was oxidized by $H_2O_2$, and the oxidation product of TMB was readily measured by a UV–Vis spectrophotometer or even by the naked eye. In the absence of the target, the biotin—antibody, SA, and linker DNA were washed away. Without the linker DNA, the hairpin structures of Bio-H1 and Bio-H2 were not able to open and assemble into a long double-helix, resulting in a low absorption peak. Thus, the concentrations of the target protein were determined simply by monitoring the intensities of the UV–Vis absorbance or observing the variation of the color by the naked eye. (Reprinted with the permission from Ref. [31]. Copyright 2015 Elsevier)

immunocomplex by streptavidin. Then, the biotinylated bovine serum albumin (Biotin-BSA) and streptavidin–horseradish peroxidase (SA-HRP) assembled at a node of the assembled products of nucleic acids due to the biotin–streptavidin reaction. Under the action of the catalysis of horseradish peroxidase, TMB was oxidized by $H_2O_2$ and the consequent product could be measured by its UV–Vis absorbance signal and sensitive colorimetric detection. This proposed colorimetric method showed a wide linear range from 5 to 1000 pg/mL with the limit of detection (LOD) of 1.95 pg/mL by the instrument, and even 5 pg/mL target protein could be easily distinguished by naked eyes.

## 2.2.2 Open Sandwich Immunoassays (OS-IA)

Although colorimetry sandwich assays have been developed very well, immuno-logical detection of small haptens or peptides is usually performed primarily in the form of competition, which requires corresponding competitor molecules labeled with signal reporters. In order to reduce the labeling of small molecules and improve the detection requirements for low molecular weight substances, Ueda et al. developed an open sandwich immunoassay (OS-IA) on the basis of antigen-dependent stabilization of the antibody variable region to quantify various antigens and for the noncompetitive detection of small molecules [32]. As shown in Fig. 2.3, OS-IA on the concept of building in the interaction between two antibody variable domains (VH and VL) is dependent on the presence of antigen. Without antigens, many fragments in the two V region fragments are easy to dissociate, while in the presence of antigens, these V region fragments turn to the bridging antigen and form a sandwich structure. The enzyme that is labeled will catalyze the addition of the substrate material, resulting in a change in the color of the system, indicating whether the antigen is present and its concentration. OS-IA has several advantages over conventional sandwich immunoassays, such as the detection of monovalent antigens, wider working range, shorter measurement time, and availability for homogeneous immunoassay [33, 34]. Ueda et al. has achieved the detection limits of 0.1 ng/mL by employing OS-IA to the detection of thyroxine (T4), which is the first report in regard to the selection for an anti-hapten antibody with no hapten–carrier conjugates. This method should be especially suitable for selecting antibody fragments which have better performance in hapten OS-IA [35].

### 2.2.3 Gold–Multienzyme–Nanocarrier-Based Colorimetric Sandwich Assays

In the past few years, nanomaterial disciplines have developed rapidly. Due to the high surface area-to-volume ratio of nanomaterials, the ability of multienzyme carrier can be realized. Therefore, the sensitivity of nanomaterials based on the sensitivity of protein colorimetric sandwich assays is greatly improved [36, 37]. It is worth noting that AuNPs (Fig. 2.3) have several advantages as described above, which are currently the most practical multienzyme carriers for the immobilization of multienzymes [9–11]. Gold–multienzyme–nanocarrier-based colorimetric sandwich assays were developed for the detection of CA15-3 antigen, which is a significant biomarker in the blood for follow-up treatment of breast cancer. AuNPs were employed as carriers of the signaling antibody anti-CA15-3-HRP in order to obtain an amplification of the optical signal [38]. The schematic diagram of the principle and steps of gold–multienzyme–nanocarrier-based colorimetric sandwich assays is shown in Fig. 2.4. Compared with the classical ELISA method, this assay achieved higher sensitivity and shorter measurement time. Hewlett et al. also assessed the feasibility of using gold–multienzyme–nanocarrier-based colorimetric sandwich assays for enhancing the detection sensitivity of HIV-1 capsid (p24) antigen [39]. Chen et al. proposed a plasmonic ELISA on the basis of the highly sensitive colorimetric detection of ALP that is carried out by iodine-mediated

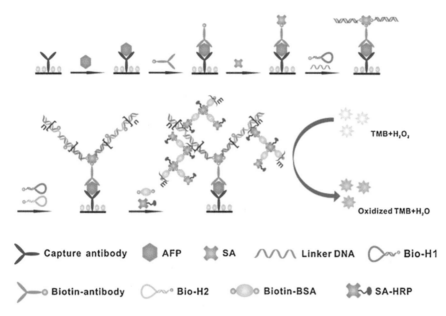

**Capture antibody**   **AFP**   **SA**   **Linker DNA**   **Bio-H1**

**Biotin-antibody**   **Bio-H2**   **Biotin-BSA**   **SA-HRP**

**Fig. 2.3** Schematic illustration of the principle of OS-ELISA. When in the presence of T4 binders, the variable domains of VL and VH turn to the bridging T4 and form a sandwich structure. The enzyme that is labeled will catalyze the addition of the substrate material, resulting in a change in the color of the system, indicating whether the antigen is present and its concentration (Reprinted with the permission from Ref. [32]. Copyright 2011 American chemical society)

etching of AuNPs [40]. This method has obtained an ultralow LOD (i.e., 100 pg/ mL) for human immunoglobulin G (IgG), and more importantly, the visual LOD (3.0 ng/mL) allows for differential diagnosis with the help of naked eyes. In addition, the gold–multienzyme–nanocarrier-based colorimetric sandwich assay was also successfully used for the detection of fetal globulin A/AHSG [41]. In theory, an amplified detection limit can be lowered to about 0.1 pg/mL by Au nanoparticles with superior carrying capability, which is about 100–150 times lower than the 1:1 enzyme-antibody conjugation method. The above results demonstrate that the NP-based common labeling technology and its application can offer a fast and sensitive test platform for laboratory research and clinical diagnosis.

## 2.3   Colorimetric Assays Based on Biomimetic Nanomaterials

Nanomaterial-based enzymes have many advantages over natural enzymes, such as stable in a wide range of pH, temperature, and inertia of several kinds of protease [13–17]. As shown in Fig. 2.5, several kinds of nanoparticles, including Pt nanoparticles, Au nanorods coated with a shell composed of Pt nanodots (Au@Pt nanostructures), $Fe_3O_4$ magnetic nanoparticles, FeS nanosheets, graphene oxide

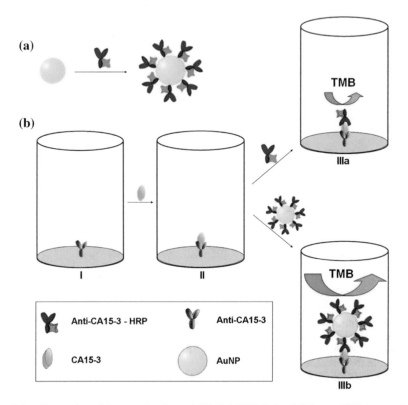

**Fig. 2.4** **a** Preparation of the complex Au-anti-CA15-3-HRP. **b** Sandwich-type ELISA procedure without (IIIa) and with (IIIb) the application of Au NPs as the signal enhancer. The capture anti-CA15-3 antibody was first immobilized on 96-well microplate and blocked with BSA. Then, CA15-3 antigen with different concentrations was added to respective wells of the microplate and incubated for 1 h. After that, the wells were washed several times and HRP-labeled detecting antibody unconjugated or conjugated to AuNPs was added and incubated for 1 h. TMB enzyme substrate is added to each well and incubated on a plate shake, and color development was stopped by introducing of sulfuric acid. The optical density was read at 450 nm with a microplate reader within 15 min after stopping the reaction. (Reprinted with the permission from Ref. [38]. Copyright 2010 American chemical society)

nanoparticles, single-wall carbon nanotubes, and cupric oxide nanoparticles, have been found to show novel performance with horseradish peroxidase (HRP) and exhibit great potential in colorimetric sandwich assays in recent years.

## 2.3.1 Fe₃O₄ Magnetic Nanoparticles (MNPs) Biomimetic Enzymes

Among these nanostructured biomimetic enzymes, the intrinsic peroxidase-like properties and magnetic properties facilitate separation of $Fe_3O_4$ magnetic

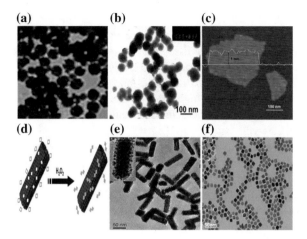

**Fig. 2.5** Various nanomaterials that were employed as nanomaterial-based enzymes. **a** Peroxidase-like $Fe_3O_4$ nanoparticles (Reprinted with the permission from Ref. [13]. Copyright 2007 Nature Publishing Group). **b** FeS nanoparticles (Reprinted with the permission from Ref. [16]. Copyright 2009 Wiley-VCH Verlag GmbH & Co. KGaA). **c** Graphene oxide (Reprinted with the permission from Ref. [42]. Copyright 2010 Wiley-VCH Verlag Copyright 2010 Wiley-VCH Verlag GmbH & Co. KGaA). **d** Single-wall carbon nanotubes (Reprinted with the permission from Ref. [15]. Copyright 2010 Wiley-VCH Verlag GmbH & Co. KgaA). **e** Au@Pt nanostructures (Reprinted with the permission from Ref. [43]. Copyright 2011 Elsevier). **f** Pt nanoparticles (Reprinted with the permission from Ref. [44]. Copyright 2011 Elsevier)

nanoparticles (MNPs), so it has gained much attention [13, 45]. Gao et al. demonstrated that MNP biomimetic enzymes were superior to traditional enzymes in several ways: (a) Peroxidase activity was preserved in a wide range of pH and temperature ranges; (b) they have intrinsic dual functionality, as a peroxidase and magnetic separator; (c) they have size-dependent catalytic properties—the smaller the size, the higher the catalytic activity; (d) the synthesis for $Fe_3O_4$ MNPs has many advantages such as easy preparation, low cost, high activity, and economical. For example, Song et al. developed chitosan-modified $Fe_3O_4$ MNPs that were utilized to mimic peroxidase for the detection of thrombin over a linear range from 1 to 100 nM with LOD lowered to 1 nM [45].

## 2.3.2 Au@Pt Nanostructure Biomimetic Enzymes

Au@Pt nanostructures showed intrinsic oxidase-like, peroxidase-like, and catalase-like activity, catalyzing oxygen and $H_2O_2$ reduction, and the dismutation decomposition of $H_2O_2$ to produce oxygen [43]. Tang et al. utilized Au@Pt nanostructures as efficient peroxidase biomimetic nanomaterials for enhanced signal of calorimetric immunoassay of prostate-specific antigen in situ [46]. The results show Au@Pt nanostructures based on colorimetric immunoassays could have effective response toward prostate-specific antigen in the working range of 5–500 pg/mL with low LOD of 2.9 pg/mL. Guo et al. developed an Au@Pt nanostructures based on ELISA for the detection of mouse interleukin 2 (IL-2), and the lowest limit of detection was estimated to be 1 pg/mL [43]. Chen et al. employed an aptamer-based sandwich assay to design a label-free colorimetric aptasensor for the detection of protein with high sensitivity and selectivity by synthesizing DNA-templated Ag/Pt nanostructures, which possess highly efficient peroxidase-like catalytic activity [47]. In general, compared to HRP, Au@Pt nanostructures have advantages such as easy preparation, low cost, better stability, and tunable catalytic activity, which makes them a promising enzyme mimetic candidate and may find potential applications in bioassays, biocatalysis, and nanobiomedicine like reactive oxygen species (ROS)-related fields (anti-aging and therapeutics for neurodegenerative diseases and cancers) (Fig. 2.6).

Nanostructured enzyme mimics own several merits over traditional natural enzymes, such as easy preparation, robustness, stability in rough conditions, tunable catalytic properties, and multiple functions, which in turn enhance the detection. These nanostructured enzyme-mimic-based sandwich colorimetric assays may have potential applications to the detection not only of protein but also of many other

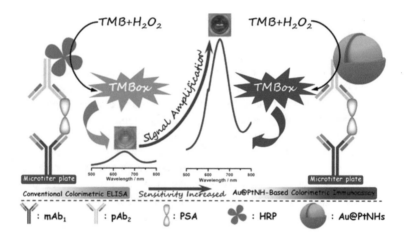

**Fig. 2.6** Schematic illustration of conventional colorimetric enzyme-linked immunosorbent assay (left) and urchin-like (gold core)@(platinum shell) nanohybrid-based colorimetric immunoassay (right). mAb1 antibody was first immobilized on a high-binding microplate by introducing of mAb1, and then, the microplate was washed with washing buffer and incubated again with blocking buffer to decrease the nonspecific adsorption. Following that, the detection of target PSA with a sandwich-like immunoassay mode by the mAb1-functionalized microplate is as follows: (i) PSA standard sample was added to form the antigen–antibody immunocomplex; (ii) after washing, Au@PtNH-pAb₂ suspension was injected into the plate and forms a sandwiched-like immune complex; (iii) after washing, TMB and $H_2O_2$ were thrown into the microplate and incubated to produce colored product; and (iv) the absorbance was collected, and the colorimetric signal at 650 nm is recorded using the microplate reader. (Reprinted with the permission from Ref. [46].Copyright 2015 Elsevier)

biomolecules. These results provide a novel idea for the development of highly sensitive, highly stable, and inexpensive nonenzyme immunoassay platforms, which acts as an alternative to conventional enzyme-based immunoassay platforms.

## 2.4 Conclusion

In this chapter, we focus on the recent development of several biosensors based on colorimetric sandwich assay and presented design strategies and typical examples for the detection of proteins. The colorimetric assay is simpler, and their response can be detected by photometry or with bare eyes due to the intrinsic good sensitivity and simplicity; usually, a long analysis time takes up to 2 h. Moreover, the colorimetric sandwich assay based on biomimetic nanomaterials has several advantages over traditional enzymes, such as easy preparation, stability under rough conditions, adjustable catalytic performance, and multiple functions. This has improved the detection sensitivity and efficiency of colorimetric sandwich assays. With the progress of chemistry, biotechnology, and nanotechnology, the determination of color sandwich has been widely developed, and we hope that the colorimetric sandwich assay will has more and more application to many hospitals and diagnostic sites.

## References

1. Liu YX, Huang XG, Ren JC (2016) Recent advances in chemiluminescence detection coupled with capillary electrophoresis and microchip capillary electrophoresis. Electrophoresis 37:2–18
2. Klapkova E, Fortova M, Prusa R, Moravcova L, Kotaska K (2016) Determination of urine albumin by new simple high-performanceliquid chromatography method. J Clin Lab Anal 30:1226–1231
3. Doussineau T, Mathevon C, Altamura L, Vendrely C, Dugourd P, Forge V, Antoine R (2016) Mass determination of entire amyloid fibrils by using mass spectrometry. Angew Chem Int Ed 55:2340–2344
4. Engvall E, Perlmann P (1971) Enzyme-linked immunosorbent assay (ELISA). Quant Assay Immunoglobulin G. Immunochem 8:871–874
5. Tomita H, Ogawa M, Kamijo T, Mori O, Ishikawa E, Mohri Z, Murakami Y (1989) A highly sensitive sandwich enzyme immunoassay of urinary growth hormone in children with short stature, Turner's syndrome, and simple obesity. Acta Endocrinol 121:513–519
6. Li W, Qiang WB, Li J, Li H, Dong YF, Zhao YJ, Xu DK (2014) Nanoparticle-catalyzed reductive bleaching for fabricating turn-off and enzyme-free amplified colorimetric bioassays. Biosens Bioelectron 51:219–224
7. Arya SK, Bhansali S (2011) Lung cancer and its early detection using biomarker-based biosensors. Chem Rev 111:6783–6809
8. Lei JQ, Jing T, Zhou TT, Zhou YS, Wu W, Mei SR, Zhou YK (2014) A simple and sensitive immunoassay for the determination of human chorionic gonadotropin by graphene-based chemiluminescence resonance energy transfer. Biosens Bioelectron 54:72–77

9. Ke RQ, Yang W, Xia XH, Xu Y, Li QG (2010) Tandem conjugation of enzyme and antibody on silica nanoparticle for enzyme immunoassay. Anal Biochem 406:8–13

10. Ambrosi A, Castaneda MT, Killard AJ, Smyth MR, Alegret S, Merkoci A (2007) Double-codified gold nanolabels for enhanced immunoanalysis. Anal Chem 79:5232–5240

11. Liu MY, Jia CP, Huang YY, Lou XH, Yao SH, Jin QH, Zhao JL, Xiang JQ (2010) Highly sensitive protein detection using enzyme-labeled gold nanoparticle probes. Analyst 135:327–331

12. Shen JW, Li YB, Gu HS, Xia F, Zuo XL (2014) Recent development of sandwich assay based on the nanobiotechnologies for proteins, nucleic acids, small molecules, and Ions. Chem Rev 114:7631–7677

13. Gao LZ, Zhuang J, Nie L, Zhang JB, Zhang Y, Gu N, Wang TH, Feng J, Yang DL, Perrett S, Yan XY (2007) Intrinsic peroxidase-like activity of ferromagnetic nanoparticles. Nat Nanotechnol 2:577–583

14. Wei H, Wang E (2013) Nanomaterials with enzyme-like characteristics (nanozymes): next-generation artificial enzymes. Chem Soc Rev 42:6060–6093

15. Song YJ, Wang XH, Zhao C, Qu KG, Ren JS, Qu XG (2010) Label-free colorimetric detection of single nucleotide polymorphism by using single-walled carbon nanotube intrinsic peroxidase-like activity. Chem—Eur J 16:3617–3621

16. Dai ZH, Liu SH, Bao JC, Jui HC (2009) Nanostructured FeS as a mimic peroxidase for biocatalysis and biosensing. Chem—Eur J 15:4321–4326

17. Chen W, Chen J, Liu AL, Wang LM, Li GW, Lin XH (2011) Peroxidase-like activity of cupric oxide nanoparticle. ChemCatChem 3:1151–1154

18. Cleland WW, Hengge AC (2006) Enzymatic mechanisms of phosphate and sulfate transfer. Chem Rev 106:3252–3278

19. Ko YC, Mukaida N, Panyutich A, Voitenok NN, Matsushima K, Kawai T, Kasahara T (1992) A sensitive enzyme-linked-immunosorbent-assay for human interleukin-8. J Immunol Methods 149:227–235

20. Daniilidou M, Tsolaki M, Giannakouros T, Nikolakaki E (2011) Detection of elevated antibodies against SR protein kinase 1 in the serum of Alzheimer's disease patients. J Neuroimmunol 238:67–72

21. Vandermeeren M, Mercken M, Vanmechelen E, Six J, Vandevoorde A, Martin JJ, Cras P (1993) Detection of tau proteins in normal and alzheimers-disease cerebrospinal-fluid with a sensitive sandwich enzyme-linked-immunosorbent-assay. J Neurochem 61:1828–1834

22. Surojanametakul V, Doi H, Shibata H, Mizumura T, Takahashi T, Varanyanond W, Wannapinpong S, Shoji M, Ito T, Tamura H (2011) Reliable enzyme-linked immunosorbent assay for the determination of coconut milk proteins in processed foods. J Agric Food Chem 59:2131–2136

23. Mercken M, Vandermeeren M, Lubke U, Six J, Boons J, Vanmechelen E, Vandevoorde A, Gheuens J (1992) Affinity purification of human tau-proteins and the construction of a sensitive enzyme-linked-immunosorbent-assay for tau-detection. J Neurochem 58:548–553

24. Frey A, Meckelein B, Externest D, Schmidt MA (2000) A stable and highly sensitive 3, 3', 5, 5'-tetramethylbenzidine-based substrate reagent for enzyme-linked immunosorbent assays. J Immunol Methods 233:47–56

25. El Mouedden M, Vandermeeren M, Meert T, Mercken M (2005) Development of a specific ELISA for the quantitative study of amino-terminally truncated beta-amyloid peptides. J Neurosci Methods 145:97–105

26. Castigliego L, Li XN, Armani A, Mazzi M, Guidi A (2011) An immunoenzymatic assay to measure insulin-like growth factor 1 (IGF-1) in buffalo milk with an IGF binding protein blocking pre-treatment of the sample. Int Dairy J 21:421–426

27. Wang J, Cao Y, Xu YY, Li GX (2009) Colorimetric multiplexed immunoassay for sequential detection of tumor markers. Biosens Bioelectron 25:532–536

28. Park JH, Cho YS, Kang S, Lee EJ, Lee GH, Hah SS (2014) A colorimetric sandwich-type assay for sensitive thrombin detection based on enzyme-linked aptamer assay. Anal Biochem 462:10–12

29. Abuknesha RA, Jeganathan F, DeGroot R, Wildeboer D, Price RG (2010) Detection of proteases using an immunochemical method with haptenylated-gelatin as a solid-phase substrate. Anal Bioanal Chem 396:2547–2558

30. Gehring A, He XH, Fratamico P, Lee J, Bagi L, Brewster J, Paoli G, He YP, Xie YP, Skinner C, Barnett C, Harris D (2014) A high-throughput, precipitating colorimetric sandwich ELISA microarray for shiga toxins. Toxins 6:1855–1872

31. Chen CH, Liu YF, Zheng ZH, Zhou GH, Ji XH, Wang HZ, He ZK (2015) A new colorimetric platform for ultrasensitive detection of protein and cancer cells based on the assembly of nucleic acids and proteins. Anal Chim Acta 880:1–7

32. Islam KN, Ihara M, Dong JH, Kasagi N, Mori T, Ueda H (2011) Direct construction of an open-sandwich enzyme immunoassay for one-step noncompetitive detection of thyroid hormone T4. Anal Chem 83:1008–1014

33. Suzuki C, Ueda H, Mahoney W, Nagamune T (2000) Open sandwich enzyme-linked immunosorbent assay for the quantitation of small haptens. Anal Biochem 286:238–246

34. Aburatani T, Sakamoto K, Masuda K, Nishi K, Ohkawa H, Nagamune T, Ueda H (2003) A general method to select antibody fragments suitable for noncompetitive detection of monovalent antigens. Anal Chem 75:4057–4064

35. Liu XB, Eichenberger M, Fujioka Y, Dong JH, Ueda H (2012) Improved detection sensitivity and selectivity attained by open-sandwich selection of an anti-estradiol antibody. Anal Sci 28:861–867

36. Wang SS, Chen ZP, Choo J, Chen LX (2016) Naked-eye sensitive ELISA-like assay based on gold-enhanced peroxidase-like immunogold activity. Anal Bioanal Chem 408:1015–1022

37. Zor E, Bekar N (2017) Lab-in-a-syringe using gold nanoparticles for rapid colorimetric chiral discrimination of enantiomers. Biosens Bioelectron 91:211–216

38. Ambrosi A, Airo F, Merkoci A (2010) Enhanced gold nanoparticle based ELISA for a breast cancer biomarker. Anal Chem 82:1151–1156

39. Tang SX, Hewlett I (2010) Nanoparticle-based immunoassays for sensitive and early detection of HIV-1 capsid (p24) antigen. J Infect Dis 201:S59–S64

40. Zhang ZY, Chen ZP, Wang SS, Cheng FB, Chen LX (2015) Iodine-mediated etching of gold nanorods for plasmonic ELISA based on colorimetric detection of alkaline phosphatase. ACS Appl Mater Interfaces 7:27639–27645

41. Dixit CK, Vashist SK, O'Neill FT, O'Reilly B, MacCraith BD, O'Kennedy R (2010) Development of a high sensitivity rapid sandwich ELISA procedure and its comparison with the conventional approach. Anal Chem 82:7049–7052

42. Song YJ, Qu KG, Zhao C, Ren JS, Qu XG (2010) Graphene oxide: intrinsic peroxidase catalytic activity and its application to glucose detection. Adv Mater 22:2206–2210

43. He WW, Liu Y, Yuan JS, Yin JL, Wu XC, Hu XN, Zhang K, Liu JB, Chen CY, Ji YL, Guo YT (2011) Au@Pt nanostructures as oxidase and peroxidase mimetics for use in immunoassays. Biomaterials 32:1139–1147

44. Ma M, Zhang Y, Gu N (2011) Peroxidase-like catalytic activity of cubic Pt nanocrystals. Colloids Surf A Physicochem Eng Asp 373:6–10

45. Song P, Wang Q, Zhang Z, Yang ZX (2010) Synthesis and gas sensing properties of biomorphic LaFeO3 hollow fibers templated from cotton. Sens Actuators, B 147:248–254

46. Gao ZQ, Xu MD, Lu MH, Chen GN, Tang DP (2015) Urchin-like (gold core)@(platinum shell) nanohybrids: a highly efficient peroxidase-mimetic system for in situ amplified colorimetric immunoassay. Biosens Bioelectron 70:194–201

47. Zheng C, Zheng AX, Liu B, Zhang XL, He Y, Li J, Yang HH, Chen GN (2014) One-pot synthesized DNA-templated Ag/Pt bimetallic nanoclusters as peroxidase mimics for colorimetric detection of thrombin. Chem Commun 50:13103–13106

# Chapter 3
# Fluorescence Sandwich Assays for Protein Detection

**Fujian Huang and Fan Xia**

**Abstract** The detection of protein with high sensitivity and selectivity is of great important for protein fundamental functions study and clinic diagnostics. Sandwich assays have been developed for multivalent proteins detection and have been prevailing for decades in the field of clinical diagnostics and bio-detection. The sandwich assays usually give a high sensitivity and selectivity because of the usage of a couple of match recognition probe and signal probe. This chapter summarizes recent advances in the sandwich assays for protein detections with fluorescence as signal outputs. Different recognition or signal elements such as antibodies and aptamers and fluorescence signal reporters (organic dyes, nanomaterials, and conjugated polymers) are discussed in details in this chapter.

**Keywords** Protein detection · Fluorescence signal reporter · Sandwich assay
Aptamer · Antibody

---

The original version of this chapter was revised: Foreword has been included and authors' affiliations have been updated. The erratum to this chapter is available at https://doi.org/10.1007/978-981-10-7835-4_13

---

F. Huang (✉) · F. Xia
Engineering Research Center of Nano-Geomaterials of Ministry of Education,
Faculty of Materials Science and Chemistry, China University of Geosciences,
Wuhan 430074, People's Republic of China
e-mail: huangfj@cug.edu.cn

F. Xia
e-mail: xiafan@cug.edu.cn; xiafan@hust.edu.cn

F. Xia
Hubei Key Laboratory of Bioinorganic Chemistry & Materia Medica,
School of Chemistry and Chemical Engineering, Huazhong University
of Science and Technology, Wuhan 430074, People's Republic of China

© Springer Nature Singapore Pte Ltd. 2018
F. Xia et al. (eds.), *Biosensors Based on Sandwich Assays*,
https://doi.org/10.1007/978-981-10-7835-4_3

29

## 3.1  Introduction

Proteins are biotechnological products which could be biomarkers for diseases, health states, and other adverse effects [1, 2]. Thus, sensitive and specific detection of target proteins is highly desirable for clinical diagnosis, therapy monitoring, and environmental monitoring [3]. For decades, the sandwich assays especially fluorescence sandwich assays have been widely used and prevailing in the detection of target proteins. These sandwich assays usually rely on the simultaneous binding of the two recognition molecules with target proteins, which makes these detection assays extremely specific even in complex samples. In addition, because of the high affinity between recognition molecules and target proteins and the usage of enzyme catalytic or amplified signaling strategies, the sandwich assays usually give satisfying detection sensitivity. Sandwich assays have the feature that the signal increases with target analyte concentration, which leads to more sensitive because the background is low. However, two specific binding pairs with spatially distant binding sites are required, which would be a problem for proteins with small size and proteins without multivalent binding sites [4].

## 3.2  Design Principles of Fluorescence Sandwich Assays

Sandwich assays, as the name implies, are theoretically composed of three basic elements, that is, target analytes, recognition units, and the signal marker [5]. These assays named as sandwich assays because the target analytes are sandwiched between two different recognition units. In sandwich assays, the target protein is captured between two recognition (binding) pairs, of which one is usually immobilized on the solid surface and the other labeled with signal markers. Figure 3.1 shows a typical fluorescence sandwich assays for protein detection. As can be seen, the recognition units are usually composed of two different molecules: recognition molecules 1 (recognition probe) and recognition molecules 2 (signal probe). In a typical detection process, the specific recognition of target protein was initially performed through specific binding between target analytes and recognition molecules 1. After that, the recognition molecules 2 then specifically interact with other site of target protein to introduce the fluorescence signal marker. Usually, the fluorescence signal marker attached to the recognition molecules 2, so that, in response or upon to binding of the target protein to the recognition molecules 2, the signal marker outputs a readily detectable fluorescence signal, providing the assay with a sensitive measurable fluorescence to indicate the presence and concentration of the target protein. In theory, the fluorescence sandwich can provide the highest level of sensitivity and specificity because of the use of a couple of recognition molecules and the high affinity between recognition molecules and target proteins.

As mentioned above, the major process for the protein detection by using sandwich assays is target proteins sandwiching process between recognition probe

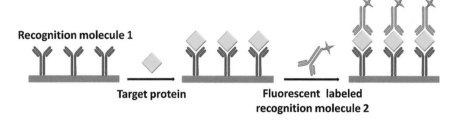

**Fig. 3.1** Scheme of fluorescence sandwich assays for protein detection. In a typical sandwich assay, the recognition molecule 1 was usually immobilized on the solid support surface, target proteins are then captured by the immobilized recognition molecule 1. The second recognition molecule (recognition molecule 2), with fluorescent labeling, binds the captured target protein on the other bind site to generate the readable fluorescence signal

and signal probe modified with the signaling marker. Different recognition molecules, such as antibodies and aptamers, could be used to specifically and sensitively recognize target proteins. At the same time, with the progress in chemistry, the signal probe can be feasibly modified with various signal markers, such as fluorescein, nanomaterials, radionuclides, redox tags, and enzymes, which in turn allows for a broader variety of readouts in a sandwich assay. This chapter focuses on the recent developments of sandwich assays for protein by using antibodies and aptamers as recognition units and with fluorescence as signal outputs. The commonly used fluorescence signal markers including traditional organic dyes, fluorescence nano-materials, and fluorescent conjugated polymers are also discussed.

## 3.3  Antibodies as Recognition Probe or Signal Probe

Antibodies are a large, usually Y-shaped proteins, which recognizes a unique molecule of the antigen with high specificity and affinity. By using antibodies as recognition probe or signal probe, a kind of sandwich assays termed as immunoassay can be fabricated for protein detection [6]. An immunoassay is a biochemical analysis that measures the presence or concentration of a target analyte in a solution by using an antibody. Immunoassays for protein detection rely on the ability of an antibody to recognize and bind a target protein [7]. In addition to the binding of an antibody to its target protein, the other key feature of all immunoassays is a means to produce a measurable signal in response to the binding [8]. This requires the usage of second antibody which usually chemically linked with some kind of detectable label. Immunoassays employ a variety of different labels, such as enzymes, radioactive isotopes, DNA reporters, electrochemilumi-nescent tags, and fluorescent reporters, to allow for detection of target proteins. The concept of fluorescence immunoassay was first proposed to detect pneumococcus by Albert H. Coons in the early 1940s, and till now fluorescence immunoassay has been prevailing both in clinical diagnostics and in research because of its high

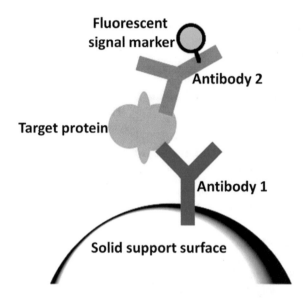

**Fig. 3.2** Schematic illustration of a typical structure of sandwich-type immunoassays for target protein detections with fluorescence as output signal. Antibody 1 is immobilized on the solid surface, and target proteins are then captured by the immobilized antibody. The second antibody (antibody 2), with fluorescent labeling, binds the captured target protein on the other bind site to generate the readable fluorescence signal

sensitivity and specificity [5]. Figure 3.2 shows a typical structure of sandwich-type immunoassays for protein detection with fluorescence as output signal. Antibody 1 is immobilized on the solid surface (such as 96 well plate and microspheres); target proteins are then captured by the immobilized antibody. The second antibody (antibody 2), with fluorescent labeling, binds the captured target protein on the other bind site to generate the readable fluorescence signal.

Finding suitable antibody pairs for target proteins is the most important step in developing antibody-based sandwich assays. It is extremely challenging and time-consuming to generate monoclonal antibody pairs because antibodies are usually generated toward the most immunogenic domain on the target protein (often called the "hot spot"), while other antibodies for the antibody pairs are difficult to obtain [9, 10]. Thus, polyclonal antibodies pairs are often used in sandwich assays and only a few monoclonal antibody pairs are available for sandwich assays [11, 12]. In order to extend the application scope of sandwich assays, researchers are trying to use alternative recognition units instead of antibody pairs [13–15].

## 3.4 Aptamers as Recognition Probe or Signal Probe

Different with antibodies, nucleic acid aptamers are artificial DNA/RNA oligonucleotide that bind to a specific target molecule [16, 17]. Aptamers are usually
created by selecting them from a large random sequence pool through repeated
rounds of in vitro selection called systematic evolution of ligands by exponential
enrichment (SELEX) to bind to various molecular targets such as small molecules
[18], proteins [19], and even cells [20, 21]. Similar with antibody, aptamers show a
high binging specificity and affinity to the target proteins. This character makes
aptamer suitable as biosensoring elements for protein detection. Besides, aptamers
have the features including synthetic convenience, cost-effectiveness, flexible
modification, easy regeneration capabilities, and thermostability [22–24]. In these
senses, aptamers as recognition elements for protein detection show competitive
advantages over antibodies. To date, aptamers as recognition units have been
widely used in the field of biosensing for their promising potential in detection
different kinds of targets with method including fluorescence [25, 26], colorimetry
[27, 28], chemiluminescence [29], surface plasmon resonance (SPR) [30, 31], and
electrochemistry [32, 33]. Most of the aptamer-based biosensors are created on the
basis of target-binding responsive conformational changes of aptamers [34].
However, aptamer conformational changes are sometimes susceptible to interfering
factors in complex matrixes, which may result in nonspecific and false signals and
thus limit the usage of aptamers for complex samples detection [35].

To improve the specificity, dual-aptamer binding scheme or equivalently,
sandwich assays have been developed for protein detection [36]. The sandwich
assay is fairly satisfactory for protein detection because of its high sensitivity and
specificity due to the dual recognition mechanism [4]. In aptamer-based sandwich
assay, two different aptamer probes are simultaneously required for both target
protein capture and signal generation [37]. These two aptamers usually recognize
two spatially distant domains of a target protein, resulting in the improved specificity (Fig. 3.3) [38]. The sandwich-type configuration can improve the detection
specificity, however, it required the target proteins have two different aptamer
binding sites. This requirement makes the aptamer-based sandwich assays only
suitable for multivalent protein (protein with two or more binding sites) detection.
Proteins with a small size or with only one aptamer binding site cannot be measured
by aptamer-based sandwich assays. The main reason is that a small protein usually
presents only one epitope and that steric hindrance prevents it from binding to two
probes simultaneously. Thus, only a few target proteins with multivalent aptamer
binding sites, such as thrombin [39, 40], prion protein [41], antitoxoplasma IgG
[42], and platelet-derived growth factor (PDGF) [43], have been used as model
proteins to demonstrate the detection performance of aptamer-based sandwich
assays. To address this issue, a new SELEX strategy has been developed by Soh
and his coworkers for "aptamer pairs" screening [44]. By using this strategy, a
high-quality DNA "aptamer pair" for plasminogen activator inhibitor 1 (PAI-1) was
successfully isolated with only two rounds of screening. Besides the aptamer pairs

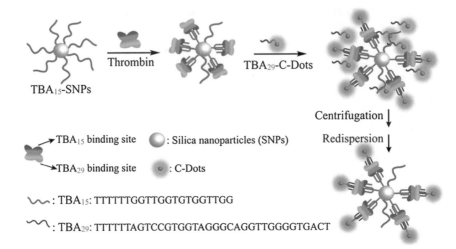

**Fig. 3.3** Schematic illustration of the fluorescence sandwich assay for thrombin detection by using two different DNA aptamers (TBA15 and TBA29) as recognition units. (Reprinted with the permission from Ref. [38]. Copyright 2012 The Royal Society of Chemistry)

screening, for the detection of target proteins without corresponding aptamers pair, the combined usage of aptamer and antibody as recognition elements would be a more direct and simple approach. By using antibody as lgE capture probe and aptamer as signal probe, a sandwich-type detection strategy has been developed for lgE detection with a detection limit down to 4.6 pM level [45].

## 3.5 Traditional Organic Dyes as Fluorescence Signal Reporter

Organic fluorophores are a kind of fluorescent chemical compound capable of absorbing light energy with a specific wavelength and re-emitting light with a longer wavelength as fluorescence signal for fluorescence sandwich assays. Traditionally, organic fluorophores including fluorescein isothiocyanate (FITC) [46], rhodamine [40], cyanine dyes [47], and Alexa [48] are usually chosen as signaling marker in fluorescence sandwich assays because of their definite emission wavelength, satisfying brightness, and commercial availability. In a typical fluorescence sandwich assay for protein detection, the detection of fluorescence signal was usually performed after additional washing to eliminate the background interference. Microbeads [49], 96 well plate [50], and microfluidic chips [51, 52] are usually used as solid support in solid-phase fluorescence sandwich assay. Microbeads, especially magnetic microspheres [37, 53], are preferred in the applications of solid-phase sandwich assays because of their simple magnetic or

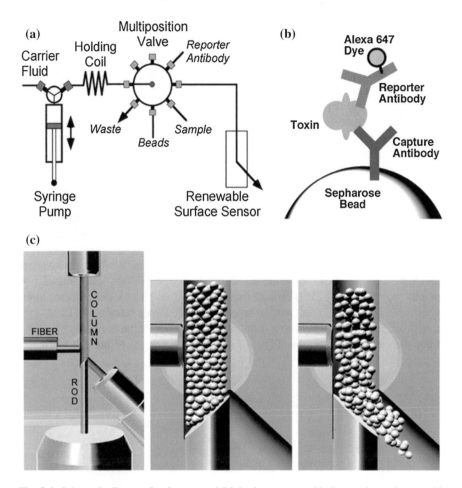

**Fig. 3.4** Schematic diagram for the sequential injection system with the rotating rod renewable surface sensor flow cell as the separator/reactor/detector (**a**), composition of the sandwich assay complex for botulinum toxin detection (**b**), and the rotating rod flow cell (**c**) (Reprinted with the permission from Ref. [54]. Copyright 2009 The Royal Society of Chemistry)

centrifugal separation, and feasible modification. Until now, microsphere-based sandwich assays combined with flow cytometer have been widely used to detect proteins. Figure 3.4 shows a microsphere-based and renewable surface fluorescence sandwich assay using a microsphere-trapping flow channel to trap antibody-modified beads with subsequent sequential injection of the sample, washing buffer, dye-labeled antibody, and final wash solutions [54]. Optical fibers coupled to the rotating rod flow cell at a 90° angle to one another deliver excitation light from a HeNe laser (633 nm) using one fiber and collect fluorescent emission light for detection with the other. After each measurement, the used sepharose beads are released and replaced with fresh beads. The simulant toxin was detected to concentrations of 10 pM in less than 20 min. The sensitivity of this system could be improved from twofold to fourfold compared to that of traditional assays, and the procedure could be shortened to 20 min.

Parallel detection of target proteins is desirable for clinical diagnose because of the coexistence of different biomarkers in patient serum. The microfluidic purification chip (MPC) is promising for rapid, ultrasensitive, point-of-care, and multiplexed detection of proteins [37, 52]. Stern and coworkers designed a novel detection system combining an MPC and a sensing reservoir [55]. As can be seen in Fig. 3.5, multiple biomarkers were captured by MPC from physiological solutions through antigen–antibody interaction. The streptavidin-HRP complex was then bounded to the surface via biotin–streptavidin interaction. After washing and UV irradiation, the photocleavable DNA linker was cleaved, releasing the streptavidin-HRP into a pure buffer for further sensing. The devices show specific and quantitative detection of two model cancer biomarkers from a 10 μL sample of whole blood in less than 20 min.

The combination of fluorescence sandwich assays and other fluorescence detection methods such as fluorescence and scattering light cross-correlation spectroscopy (FSCCS) [56], optical trapping and two-photon excitation fluorescence method [57], and total internal reflection fluorescence imaging (TIRFM) [58] offers the direct and ultrasensitive quantification of trace amounts of target proteins in complex biological samples. Figure 3.6 shows the scheme of the direct analysis

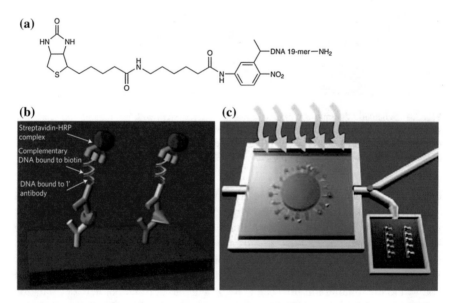

**Fig. 3.5** Principle of antiprostate-specific antigen and carbohydrate antigen 15.3 detection using a microchip-based fluorescence sandwich assays. Molecular structure of the photocleavable cross-linker (**a**). Composition of modified sandwich assays (**b**). Schematic of MPC operation (**c**). (Reprinted with the permission from Ref. [55]. Copyright 2010 Macmillan Publishers Limited)

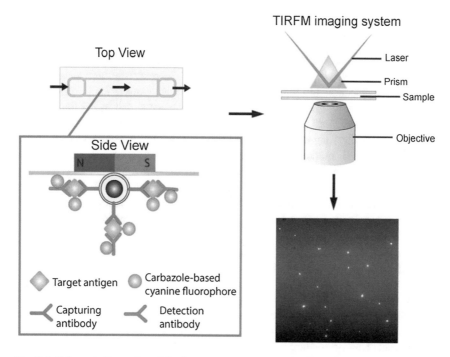

**Fig. 3.6** Schematic illustration of the fluorescence sandwich assays for carcinoembryonic antigen (CEA), prostate specific antigen (PSA), and alpha-feto-protein (AFP) detection by using total internal reflection fluorescence microscopy (TIRFM) as signal detection method. The sandwiched magnetic immuno-assembly labeled with a turn-on cyanine fluorophore was confined within the evanescent field. The fluorescence signal of each sandwiched structure confined within the evanescent was then recorded and analyzed using TIRFM. (Reprinted with the permission from Ref. [58]. Copyright 2016 The Royal Society of Chemistry)

of the single sandwiched immuno-assembly using total internal reflection fluorescence microscopy. The detection method is based on the specific antigen–antibody interactions among the target proteins, capture antibody, and detection antibody that is conjugated on the surface of magnetic nanoparticles [58]. The sandwiched immuno-assembly (MIA) is then labeled with turn-on fluorophores and analyzed with TIRFM. The TIRFM obtains high signal-to-noise images of the MIA because only MIA that are located within the evanescent field are excited by the laser. The system can successfully differentiate the target proteins from other proteins in the matrix and achieve a remarkable limit of detection down to the femto-molar regime.

## 3.6  Nanomaterials as Fluorescence Signal Reporter

Fluorescence sandwich assays are the most common approaches for target proteins detection in the field of optical biosensors due to the relatively high sensitivity of fluorescence detection and high specificity of sandwich assays. Traditional organic fluorophores labeling always give passable fluorescence signal in the fluorescence sandwich assays. Efforts to enhance the signal intensity (sensitivity) of florescence sandwich assays by integrating different kinds of novel fluorescent nanomaterials as fluorescent signal marker have attracted a great attention in recent years. Fluorescent nanostructures such as semiconductor quantum dots [50, 59, 60], carbon dots (C-Dots) [38, 61], fluorophore-doped nanoparticles [62], up-conversion nanoparticles, and other organic and inorganic fluorescence nanoparticles [63, 64] have attracted wide interest because of their unique optical and biocompatible properties.

It should be pointed out that the emerging fluorescent semiconductor quantum dots (QDs) exhibit impressive fluorescent properties such as high-quantum yields [65], long-term photostability [66], size-controlled and composition-dependent narrow emission [67], and broad adsorption [68]. These novel fluorescent properties make them an excellent candidate for the fluorescence signal marker in fluorescence sandwich assays. QDs were usually modified with antibodies or aptamers and used as fluorescent probe for target proteins detection. Cui et al. developed a versatile sandwich assay using CdTe QDs as fluorescent markers for sensitive protein detection [69]. Tu and coworkers fabricated an optical sandwich assay using CdSe/ZnS QDs as signal marker for human serum albumin detection [70]. This detection method achieved a detection limit down to 3.2 µg/mL. Warner's group developed a QD-based sandwich assays in renewable surface flow cell for rapid and sensitive fluorescence detection of botulinum neurotoxin, achieving a detection limit as low as 5 pM [50]. Liu's group proposed a versatile sandwich assay using a CdTe QD-coated silica nanoparticle as the fluorescent label for detection of rabbit IgG protein [71]. Chang and coworkers fabricated a quantum dot-based immunochromatography test strip (ICTS) for sensitive and rapid detection of alpha fetoprotein (AFP), which is a kind of biomarker for liver cancer [59]. Figure 3.7 shows the schematic illustration of the test strip and the detection process of AFP using QD-based ICTS. In a typical process, samples containing AFP was applied to the sample pad and migrated along the porous membrane by capillary action. During the migration process, the AFP combined with antibody-modified QDs in the conjugation pad to form the complexes. The formed complexes continued to move along the membrane and were captured by the other anti-AFP antibody on the test line; the excess antibody-modified QDs were captured by the secondary antibody on the control line. The fluorescence intensities of test line and control line were recorded, and the concentration of AFP was quantified according to the ratio of the intensities of test line and control line (T/C ratio). Under optimal conditions, this strip is capable of detecting as low as 1 ng/mL AFP standard analyte in 10 min with only 50 µL sample volume.

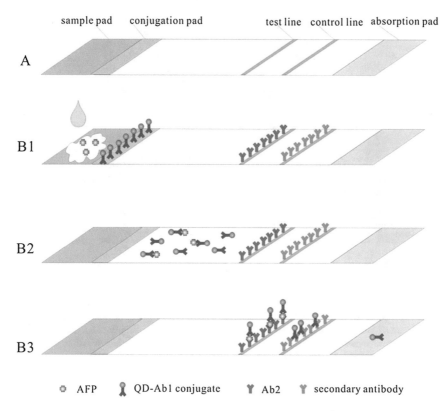

**Fig. 3.7** Schematic illustration of the principle of AFP detection by using QD-based immunochromatography test strip. (Reprinted with the permission from Ref. [59]. Copyright 2011 Elsevier)

In practical clinical diagnosis, it is desirable to develop multiplex detection approaches for simultaneous detection of a suite of biomarkers in patient's urine and serum. B. I. Swanson and his coworkers fabricated a fluorescence sandwich assay on multichannel waveguides for multiplex detection of disease-related biomarkers in complex samples by using photostable QDs as the fluorescent signal marker (Fig. 3.8) [72]. This method allows for the rapid, sensitive, and specific quantification of protective antigen and lethal factor in serum with a detection limit of 1 pM.

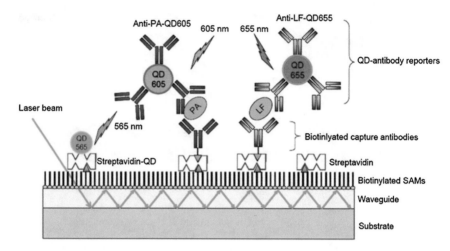

**Fig. 3.8** Scheme principle of the fluorescence sandwich assay constructed on a functionalized waveguide channel surface for multiplex target protein detection (Reprinted with the permission from Ref. [72]. Copyright 2010 American Chemical Society). The surface is functionalized with SAMs and biotinylated anti-PA and anti-LF capture antibodies, entrapped by biotin-avidin chemistry. Some of the streptavidin is labeled with QD565, using as an internal standard. Subsequent addition of the sample results in antigen (PA and LF) binding. Finally, the QD-labeled fluorescence reporters (anti-PA-QD605 and anti-LF-QD655) are added. Excitation at 535 nm results in differential emission of QDs, measured using the spectrometer interface

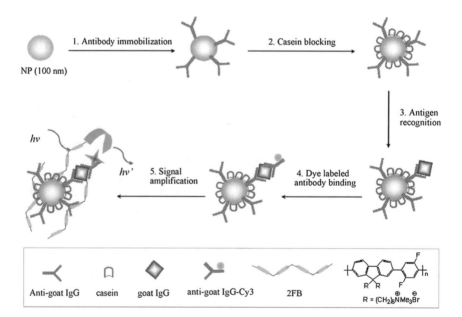

**Fig. 3.9** Schematic illustration of the CPs amplified fluorescence sandwich assay (Reprinted with the permission from Ref. [78]. Copyright 2009 Elsevier). In this assay, the fluorescence signal of organic dye was enhanced by conjugated polymers through fluorescence resonance energy transfer (FRET)

## 3.7 Conjugated Polymers as Fluorescence Signaling Reporter

Water-soluble conjugated polymers (CPs), due to their novel optical properties and photoelectric properties, have been widely used for biomolecule detections [73–75]. A novel signal amplification method via the fluorescence resonance energy transfer (FRET) between organic fluorophores and CPs has been wildly used in the field of bioanalysis [76, 77]. Incorporating this signal amplification method, a kind of modified fluorescence sandwich assay was developed by Liu and his coworkers for sensitive IgG detection (Fig. 3.9) [78]. In this assay, the fluorescence signal of organic dye was enhanced by conjugated polymers through FRET and a detection limit as low as 1.1 ng/mL was achieved.

## 3.8 Conclusion

In this chapter, we summarized the fluorescence sandwich assays for protein detection in the last decade. Fluorescence sandwich assays for proteins detections with high specificity and sensitivity have become an important tool in bioanalytical studies and clinic diagnostics. The advantages of the sandwich assays are attributed to the usage of a couple of "recognition pairs" so that the fluorescence signal is generated only when both recognition units bind simultaneously to different domains on the target proteins, resulting in a highly specific and sensitive detection in complex clinic samples. "Aptamer pairs" and "antibody pairs" have been widely used for recognition units in the fluorescence sandwich assays for protein detections due to their high selectivity and binding affinity to target proteins. Fluorescence labels such as organic fluorophore, inorganic fluorescent nanomaterials are usually used as fluorescence signal markers. The main challenge efforts to improve the performance of fluorescence sandwich assay are mainly focused on the screening of new recognition pairs for target proteins to improve the specificity [42, 44], finding novel fluorescence signal markers with high brightness or incorporating novel signal amplification methods [78–81] to improve the sensitivity to make it suitable for target protein detection in real sample, such as in patient whole blood sample, without any sample pretreatment.

## References

1. Vasan RS (2006) Biomarkers of cardiovascular disease—molecular basis and practical considerations. Circulation 113:2335–2362
2. Molitoris BA, Melnikov VY, Okusa MD, Hirnmelfarb J (2008) Technology insight: biomarker development in acute kidney injury—what can we anticipate? Nat Clin Pract Nephr 4:154–165

3. Bilitewski U (2006) Protein-sensing assay formats and devices. Anal Chim Acta 568:232–247
4. Chen AL, Yan MM, Yang SM (2016) Split aptamers and their applications in sandwich aptasensors. TrAC Trends Anal Chem 80:581–593
5. Shen JW, Li YB, Gu HS, Xia F, Zuo XL (2014) Recent development of sandwich assay based on the nanobiotechnologies for proteins, nucleic acids, small molecules, and ions. Chem Rev 114:7631–7677
6. Shankar G, Devanarayan V, Amaravadi L, Barrett YC, Bowsher R, Finco-Kent D, Fiscella M, Gorovits B, Kirschner S, Moxness M, Parish T, Quarmby V, Smith H, Smith W, Zuckerman LA, Koren E (2008) Recommendations for the validation of immunoassays used for detection of host antibodies against biotechnology products. J Pharm Biomed Anal 48:1267–1281
7. Morgan CL, Newman DJ, Price CP (1996) Immunosensors: technology and opportunities in laboratory medicine. Clin Chem 42:193–209
8. Hock B (1997) Antibodies for immunosensors—a review. Anal Chim Acta 347:177–186
9. Vanderlugt CL, Miller SD (2002) Epitope spreading in immunemediated diseases: implications for immunotherapy. Nat Rev Immunol 2:85–95
10. Kiening M, Niessner R, Weller MG (2005) Microplate-based screening methods for the efficient development of sandwich immunoassays. Analyst 130:1580–1588
11. Stoevesandt O, Taussig MJ (2007) Affinity reagent resources for human proteome detection: Initiatives and perspectives. Proteomics 7:2738–2750
12. Anderson NL, Anderson NG, Pearson TW, Borchers CH, Paulovich AG, Patterson SD, Gillette M, Aebersold R, Carr SA (2009) A human proteome detection and quantitation project. Mol Cell Proteomics 8:883–886
13. Toh SY, Citartan M, Gopinath SCB, Tang T-H (2015) Aptamers as a replacement for antibodies in enzyme-linked immunosorbent assay. Biosens Bioelectron 64:392–403
14. Jayasena SD (1999) Aptamers: an emerging class of molecules that rival antibodies in diagnostics. Clin Chem 45:1628–1650
15. Lofblom J, Feldwisch J, Tolmachev V, Carlsson J, Stahl S, Frejd FY (2010) Affibody molecules: engineered proteins for therapeutic, diagnostic and biotechnological applications. FEBS Lett 584:2670–2680
16. Ellington AD, Szostak JW (1990) Invitro selection of RNA molecules that bind specific ligands. Nature 346:818–822
17. Tuerk C, Gold L (1990) Systematic evolution of ligands by exponential enrichment: RNA ligands to bacteriophage T4 DNA polymerase. Science 249:505–510
18. Kawano R, Osaki T, Sasaki H, Takinoue M, Yoshizawa S, Takeuchi S (2011) Rapid detection of a cocaine-binding aptamer using biological nanopores on a chip. J Am Chem Soc 133:8474–8477
19. Zhang HQ, Wang ZW, Li XF, Le XC (2006) Ultrasensitive detection of proteins by amplification of affinity aptamers. Angew Chem Int Ed 45:1576–1580
20. Shangguan DH, Li Y, Tang ZW, Cao ZC, Chen HW, Mallikaratchy P, Sefah K, Yang CJ, Tan WH (2006) Aptamers evolved from live cells as effective molecular probes for cancer study. Proc Natl Acad Sci U S A 103:11838–11843
21. Zhang LQ, Wan S, Jiang Y, Wang YY, Fu T, Liu QL, Cao ZJ, Qiu LP, Tan WH (2017) Molecular elucidation of disease biomarkers at the interface of chemistry and biology. J Am Chem Soc 139:2532–2540
22. Justino CIL, Freitas AC, Pereira R, Duarte AC, Santos TAPR (2015) Recent developments in recognition elements for chemical sensors and biosensors. TrAC Trends Anal Chem 68:2–17
23. Bunka DHJ, Stockley PG (2006) Aptamers come of age—at last. Nat Rev Microbiol 4: 588–596
24. Tombelli S, Minunni M, Mascini A (2005) Analytical applications of aptamers. Biosens Bioelectron 20:2424–2434
25. Li JJ, Zhong XQ, Zhang HQ, Le XC, Zhu JJ (2012) Binding-induced fluorescence turn-on assay using aptamer-functionalized silver nanocluster DNA probes. Anal Chem 84: 5170–5174

26. Tokunaga T, Namiki S, Yamada K, Imaishi T, Nonaka H, Hirose K, Sando S (2012) Cell surface-anchored fuorescent aptamer sensor enables imaging of chemical transmitter dynamics. J Am Chem Soc 134:9561–9564
27. Gopinath SCB, Lakshmipriya T, Awazu K (2014) Colorimetric detection of controlled assembly and disassembly of aptamers on unmodified gold nanoparticles. Biosens Bioelectron 51:115–123
28. Kim YS, Kim JH, Kim IA, Lee SJ, Gu MB (2011) The affinity ratio-Its pivotal role in gold nanoparticle-based competitive colorimetric aptasensor. Biosens Bioelectron 26:4058–4063
29. Li SY, Chen DY, Zhou QT, Wang W, Gao LF, Jiang J, Liang HJ, Liu YZ, Liang GL, Cui H (2014) A general chemiluminescence strategy for measuring aptamer-target binding and target concentration. Anal Chem 86:5559–5566
30. Xie LP, Yan XJ, Du YN (2014) An aptamer based wall-less LSPR array chip for label-free and high throughput detection of biomolecules. Biosens Bioelectron 53:58–64
31. Ashley J, Li SFY (2013) An aptamer based surface plasmon resonance biosensor for the detection of bovine catalase in milk. Biosens Bioelectron 48:126–131
32. Labib M, Zamay AS, Kolovskaya OS, Reshetneva IT, Zamay GS, Kibbee RJ, Sattar SA, Zamay TN, Berezovski MV (2012) Aptamer-based impedimetric sensor for bacterial typing. Anal Chem 84:8114–8117
33. Yang ZG, Kasprzyk-Hordern B, Goggins S, Frost CG, Estrela P (2015) A novel immobilization strategy for electrochemical detection of cancer biomarkers: DNA-directed immobilization of aptamer sensors for sensitive detection of prostate specific antigens. Analyst 140:2628–2633
34. Pan L, Huang Y, Wen CC, Zhao SL (2013) Label-free fluorescence probe based on structure-switching aptamer for the detection of interferon gamma. Analyst 138:6811–6816
35. Stojanovic MN, de Prada P, Landry DW (2000) Fluorescent sensors based on aptamer self-assembly. J Am Chem Soc 122:11547–11548
36. Yue QL, Shen T, Wang L, Xu SL, Li HB, Xue QW, Zhang YF, Gu XH, Zhang SQ, Liu JF (2014) A convenient sandwich assay of thrombin in biological media using nanoparticle-enhanced fluorescence polarization. Biosens Bioelectron 56:231–236
37. Tennico YH, Hutanu D, Koesdjojo MT, Bartel CM, Remcho VT (2010) On-chip aptamer-based sandwich assay for thrombin detection employing magnetic beads and quantum dots. Anal Chem 82:5591–5597
38. Xu BL, Zhao CQ, Wei WL, Ren JS, Miyoshi D, Sugimoto N, Qu XG (2012) Aptamer carbon nanodot sandwich used for fluorescent detection of protein. Analyst 137:5483–5486
39. Liu YN, Liu N, Ma XH, Li XL, Ma J, Li Y, Zhou ZJ, Gao ZX (2015) Highly specific detection of thrombin using an aptamer-based suspension array and the interaction analysis via microscale thermophoresis. Analyst 140:2762–2770
40. Roemhildt L, Pahlke C, Zoergiebel F, Braun H-G, Opitz J, Baraban L, Cuniberti G (2013) Patterned biochemical functionalization improves aptamer-based detection of unlabeled thrombin in a sandwich assay. ACS Appl Mater Interfaces 5:12029–12035
41. Xiao SJ, Hu PP, Wu XD, Zou YL, Chen LQ, Peng L, Ling J, Zhen SJ, Zhan L, Li YF, Huang CZ (2010) Sensitive discrimination and detection of prion disease-associated isoform with a dual-aptamer strategy by developing a sandwich structure of magnetic microparticles and quantum dots. Anal Chem 82:9736–9742
42. Luo Y, Liu X, Jiang TL, Liao P, Fu WL (2013) Dual-aptamer-based biosensing of toxoplasma antibody. Anal Chem 85:8354–8360
43. Ruslinda AR, Penmatsa V, Ishii Y, Tajima S, Kawarada H (2012) Highly sensitive detection of platelet-derived growth factor on a functionalized diamond surface using aptamer sandwich design. Analyst 137:1692–1697
44. Csordas AT, Jorgensen A, Wang J, Gruber E, Gong Q, Bagley ER, Nakamoto MA, Eisenstein M, Soh HT (2016) High-throughput discovery of aptamers for sandwich assays. Anal Chem 88:10842–10847

45. Peng QW, Cao ZJ, Lau C, Kai M, Lu JZ (2011) Aptamer-barcode based immunoassay for the instantaneous derivatization chemiluminescence detection of IgE coupled to magnetic beads. Analyst 136:140–147

46. Wang XQ, Ren L, Tu Q, Wang JC, Zhang YR, Li ML, Liu R, Wang JY (2011) Magnetic protein microbead-aided indirect fluoroimmunoassay for the determination of canine virus specific antibodies. Biosens Bioelectron 26:3353–3360

47. Gruber HJ, Hahn CD, Kada G, Riener CK, Harms GS, Ahrer W, Dax TG, Knaus HG (2000) Anomalous fluorescence enhancement of Cy3 and Cy3.5 versus anomalous fluorescence loss of Cy5 and Cy7 upon covalent linking to IgG and noncovalent binding to avidin. Bioconjugate Chem 11:696–704

48. Wang JY, Wang XQ, Ren L, Wang Q, Li L, Liu WM, Wan ZF, Yang LY, Sun P, Ren LL, Li ML, Wu H, Wang JF, Zhang L (2009) Conjugation of biomolecules with magnetic protein microspheres for the assay of early biomarkers associated with acute myocardial infarction. Anal Chem 81:6210–6217

49. Wang JY, Ren L, Wang XQ, Wang Q, Wan ZF, Li L, Liu WM, Wang XM, Li ML, Tong DW, Liu AJ, Shang BB (2009) Superparamagnetic microsphere-assisted fluoroimmunoassay for rapid assessment of acute myocardial infarction. Biosens Bioelectron 24:3097–3102

50. Warner MG, Grate JW, Tyler A, Ozanich RM, Miller KD, Lou J, Marks JD, Bruckner-Lea CJ (2009) Quantum dot immunoassays in renewable surface column and 96-well plate formats for the fluorescence detection of botulinum neurotoxin using high-affinity antibodies. Biosens Bioelectron 25:179–184

51. Ziegler J, Zimmermann M, Hunziker P, Delamarche E (2008) High-performance immunoassays based on through-stencil patterned antibodies and capillary systems. Anal Chem 80:1763–1769

52. Hosokawa K, Omata M, Maeda M (2007) Immunoassay on a power-free microchip with laminar flow-assisted dendritic amplification. Anal Chem 79:6000–6004

53. Zhao W, Zhang WP, Zhang ZL, He RL, Lin Y, Xie M, Wang HZ, Pang DW (2012) Robust and highly sensitive fluorescence approach for point-of-care virus detection based on immunomagnetic separation. Anal Chem 84:2358–2365

54. Grate JW, Warner MG, Ozanich RM Jr, Miller KD, Colburn HA, Dockendorff B, Antolick KC, Anheier NC Jr, Lind MA, Lou J, Marks JD, Bruckner-Lea CJ (2009) Renewable surface fluorescence sandwich immunoassay biosensor for rapid sensitive botulinum toxin detection in an automated fluidic format. Analyst 134:987–996

55. Stern E, Vacic A, Rajan NK, Criscione JM, Park J, Ilic BR, Mooney DJ, Reed MA, Fahmy TM (2010) Label-free biomarker detection from whole blood. Nat Nanotechnol 5:138–142

56. Wang JJ, Huang XY, Liu H, Dong CQ, Ren JC (2017) Fluorescence and scattering light cross correlation spectroscopy and its applications in homogeneous immunoassay. Anal Chem 89:5230–5237

57. Li CY, Cao D, Qi CB, Chen HL, Wan YT, Lin Y, Zhang ZL, Pang DW, Tang HW (2017) One-step separation-free detection of carcinoembryonic antigen in whole serum: combination of two-photon excitation fluorescence and optical trapping. Biosens Bioelectron 90:146–152

58. Ho SL, Xu D, Wong MS, Li HW (2016) Direct and multiplex quantification of protein biomarkers in serum samples using an immuno-magnetic platform. Chem Sci 7:2695–2700

59. Yang Q, Gong X, Song T, Yang J, Zhu S, Li Y, Cui Y, Li Y, Zhang B, Chang J (2011) Quantum dot-based immunochromatography test strip for rapid, quantitative and sensitive detection of alpha fetoprotein. Biosens Bioelectron 30:145–150

60. Jokerst JV, Raamanathan A, Christodoulides N, Floriano PN, Pollard AA, Simmons GW, Wong J, Gage C, Furmaga WB, Redding SW, McDevitt JT (2009) Nano-bio-chips for high performance multiplexed protein detection: determinations of cancer biomarkers in serum and saliva using quantum dot bioconjugate labels. Biosens Bioelectron 24:3622–3629

61. Wu YY, Wei P, Pengpumkiat S, Schumacher EA, Remcho VT (2015) Development of a carbon dot (C-Dot)-linked immunosorbent assay for the detection of human alpha-fetoprotein. Anal Chem 87:8510–8516
62. Yan JL, Estevez MC, Smith JE, Wang KM, He XX, Wang L, Tan WH (2007) Dye-doped nanoparticles for bioanalysis. Nano Today 2:44–50
63. Cowles CL, Zhu XS (2011) Sensitive detection of cardiac biomarker using ZnS nanoparticles as novel signal transducers. Biosens Bioelectron 30:342–346
64. Yao JJ, Han XG, Zeng S, Zhong WW (2012) Detection of femtomolar proteins by nonfluorescent ZnS nanocrystal clusters. Anal Chem 84:1645–1652
65. Michalet X, Pinaud FF, Bentolila LA, Tsay JM, Doose S, Li JJ, Sundaresan G, Wu AM, Gambhir SS, Weiss S (2005) Quantum dots for live cells, in vivo imaging, and diagnostics. Science 307:538–544
66. Walling MA, Novak JA, Shepard JRE (2009) Quantum dots for live cell and in vivo imaging. Int J Mol Sci 10:441–491
67. Regulacio MD, Han MY (2010) Composition-tunable alloyed semiconductor nanocrystals. Acc Chem Res 43:621–630
68. Chan WCW, Nie SM (1998) Quantum dot bioconjugates for ultrasensitive nonisotopic detection. Science 281:2016–2018
69. Cui RJ, Pan HC, Zhu JJ, Chen HY (2007) Versatile immunosensor using CdTe quantum dots as electrochemical and fluorescent labels. Anal Chem 79:8494–8501
70. Tu MC, Chang YT, Kang YT, Chang HY, Chang P, Yew TR (2012) A quantum dot-based optical immunosensor for human serum albumin detection. Biosens Bioelectron 34:286–290
71. Qian J, Zhang CY, Cao XD, Liu SQ (2010) Versatile immunosensor using a quantum dot coated silica nanosphere as a label for signal amplification. Anal Chem 82:6422–6429
72. Mukundan H, Xie H, Price D, Kubicek-Sutherland JZ, Grace WK, Anderson AS, Martinez JS, Hartman N, Swanson BI (2010) Quantitative multiplex detection of pathogen biomarkers on multichannel waveguides. Anal Chem 82:136–144
73. Gaylord BS, Heeger AJ, Bazan GC (2003) DNA hybridization detection with water-soluble conjugated polymers and chromophore-labeled single-stranded DNA. J Am Chem Soc 125:896–900
74. Klingstedt T, Nilsson KPR (2011) Conjugated polymers for enhanced bioimaging. Biochim Biophys Acta Gen Subj 1810:286–296
75. He F, Tang YL, Yu MH, Feng F, An LL, Sun H, Wang S, Li YL, Zhu DB, Bazan GC (2006) Quadruplex-to-duplex transition of G-rich oligonucleotides probed by cationic water-soluble conjugated polyelectrolytes. J Am Chem Soc 128:6764–6765
76. Liu B, Bazan GC (2006) Optimization of the molecular orbital energies of conjugated polymers for optical amplification of fluorescent sensors. J Am Chem Soc 128:1188–1196
77. Pu KY, Liu B (2009) Optimizing the cationic conjugated polymer-sensitized fluorescent signal of dye labeled oligonucleotide for biosensor applications. Biosens Bioelectron 24:1067–1073
78. Wang YY, Liu B (2009) Conjugated polymer as a signal amplifier for novel silica nanoparticle-based fluoroimmunoassay. Biosens Bioelectron 24:3293–3298
79. Guo LM, Hao LH, Zhao Q (2016) An aptamer assay using rolling circle amplification coupled with thrombin catalysis for protein detection. Anal Bioanal Chem 408:4715–4722
80. Lee JU, Jeong JH, Lee DS, Sim SJ (2014) Signal enhancement strategy for a micro-arrayed polydiacetylene (PDA) immunosensor using enzyme-catalyzed precipitation. Biosens Bioelectron 61:314–320
81. Niu SY, Qu LJ, Zhang Q, Lin JH (2012) Fluorescence detection of thrombin using autocatalytic strand displacement cycle reaction and a dual-aptamer DNA sandwich assay. Anal Biochem 421:362–367

# Chapter 4
# Electrochemical Sandwich Assays for Protein Detection

**Hui Li, Shaoguang Li and Fan Xia**

**Abstract** Rapid, sensitive, and selective detection of proteins biomarker plays a very important role in early diagnostics of diseases and global health. Toward this goal, numerous researchers have devoted great efforts to develop a variety of approaches for protein detections, among which electrochemical sandwich assay appears as a very promising one because their signaling mechanism between redox-active tags and electrode renders this approach to be highly sensitive and selective, rapid, miniaturizable, and cost-effective. As such, this electron communicating signal can be readily amplified by employing enzymatic catalyst reaction, metal nanoparticles, carbon-based nanomaterials, and many other strategies, in support to further improve the sensitivity of this sensing platform.

**Keywords** Electrochemical biosensor · Sandwich assay · Protein detection
Electrochemical aptasensor · Signal amplification

The original version of this chapter was revised: Foreword has been included and authors' affiliations have been updated. The erratum to this chapter is available at https://doi.org/10.1007/978-981-10-7835-4_13

H. Li (✉) · S. Li · F. Xia
Engineering Research Center of Nano-Geomaterials of Ministry of Education,
Faculty of Materials Science and Chemistry, China University of Geosciences,
Wuhan 430074, People's Republic of China
e-mail: lihui-chem@cug.edu.cn

S. Li
e-mail: lishaoguang@cug.edu.cn

F. Xia
e-mail: xiafan@cug.edu.cn; xiafan@hust.edu.cn

F. Xia
Hubei Key Laboratory of Bioinorganic Chemistry & Materia Medica,
School of Chemistry and Chemical Engineering, Huazhong University of Science
and Technology, Wuhan 430074, People's Republic of China

© Springer Nature Singapore Pte Ltd. 2018
F. Xia et al. (eds.), *Biosensors Based on Sandwich Assays*,
https://doi.org/10.1007/978-981-10-7835-4_4

47

## 4.1  Introduction

Sensitive and selective detection of proteins plays a very important role in early detection and monitoring therapy and surgery. As one of the most important protein targets, cancer biomarkers, has been identified for many types of cancer. For example, the protein prostate-specific antigen (PSA) is a serum biomarker for early diagnostics of prostate cancer, which have been approved by the American Cancer Society (ACS) [1]. Pro-inflammatory cytokine proteins are employed as a biomarker for early cancer diagnosis because of their role in disease-related inflammatory responses and in maintaining homeostasis [2]. Another example, carcinoembryonic antigen (CEA), a highly glycosylated protein, has been used as a biomarker for lung, colorectal, liver, pancreatic, and breast cancers [3]. Therefore, it is of great importance and interests to develop a method for the detection of these biomarker protein with high sensitivity and selectivity, in support for earlier stage diagnostics of cancer diseases.

Toward this goal, researchers have devoted great efforts to developing sandwich-type biosensor platform due to its superior reproducibility and impressive sensitivity. As such, a variety of detection techniques based on sandwich assay have been explored in last decades, namely radioimmunoassay, electrochemical, colorimetric, enzyme-linked immunosorbent assays (ELISA), surface plasmon resonance (SPR)-based platforms [4]. Among them, electrochemical methods are of particular interests because their signaling mechanism, relying on electronic communication between the transducer and biomolecules, provides an effective approach for interface studies and biomolecule recognitions. Their signaling mechanisms promise electrochemical approach to be sensitive, selective, rapid, miniaturizable, and cost-effective [5].

In electrochemical assay, the sandwich architecture is composed of three elements: capture elements, target protein, and probe elements. The capture elements are primary antibodies or aptamers self-assembled onto the electrode for capturing target protein, and probe elements are antibodies or aptamers pre-labeled with redox tags for signal transduction. Typically in (bio-)electrochemistry, the protein binding-induced electronic communication between electrode surface and redox tag would generate a change in either current (as amperometry), resistance (and reactance) (as impedance spectroscopy), or charge accumulation (as potentiometry) [6], the former two of which have been much more explored than the last one. Accordingly, a variety of electrochemical techniques based on these different outputs have been developed for the electrochemical measurement. Voltammetry, as an amperometric technique and the most common electrochemical method employed for protein biosensors, employs that the binding event of the analytes causes the resulting current changes while applying a varied potential. Depending on how the applied potential is applied, a variety of voltammetry techniques are derived, such as differential pulse voltammetry (DPV), cyclic voltammetry (CV), square wave voltammetry (SWV), linear sweep voltammetry (LSV), or stripping voltammetry

(SV), differential staircase, normal pulse, reverse pulse [7]. Alternatively, electrochemical impedance spectroscopy (EIS), a technique that monitors resistance change as an output, has been also employed as a label-free technique for the sensitive measurement of target analytes, but to a much less extent presumably due to its high sensitivity to non-specific binding, for selective and sensitive protein detection [7]. These techniques used for electrochemical sandwich-type assay across the time frame from 2007 to 2017 are summarized (Chart 4.1a), among which DPV and SWV are the dominant techniques, covering 52.7 and 20.1%, respectively.

Electrochemical sandwich-based biosensors for protein detection generally are based on either antibody–antigen–antibody immunoassay or aptamer-based platform (Fig. 4.1), the former of which have seen much more than the latter (80% vs. 20%) in the last decade (Year 2007–2017) (Chart 4.1b). The detection principle includes three steps: (1) primary antibodies or capture aptamers are immobilized onto an electrode surface; (2) target antigens bind to primary antibodies or capture aptamer specifically; (3) a redox tag-labeled secondary antibody or aptamer binds to a second binding site of target antigen, and the electronic signal generated by the redox label is employed for antigen quantification. In comparison with the antibody–antigen–antibody immunoassay, the aptamer-based assay employs aptamer-pair or aptamer–antibody as the recognition element (Fig. 4.1b). The redox tags used for the signal transduction are non-amplified small organic molecules (ferrocene, methylene blue, and doxorubicin), metal ions ($Cd^{2+}$, $Ag^{2+}$), and amplified redox tags such as nanoparticles and enzymes [8–10].

These two classes of sensor platform require further signal amplification to achieve the sensitive detection of ultralow protein concentration (nanomolar or even lower range). A variety of strategies have been explored for the amplification of transducing electronic signals, such as enzyme label amplification, nanoparticles including noble metal nanoparticles and quantum dots, carbon nanotube, graphene,

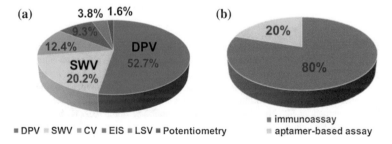

**Chart 4.1** Statistical summary of electrochemical sandwich assay for protein detection is based on the literature in the last decade (Year 2007–2017), found in ISI Web of Science using the search terms "electrochemical," "sandwich assay," and "protein detection." **a** Shown are a variety of electrochemical techniques employed for protein detection based on sandwich assay. **b** Electrochemical sandwich assay for protein detection includes two typical formats, immunoassay, and aptamer-based assay, depending on the recognition

50

H. Li et al.

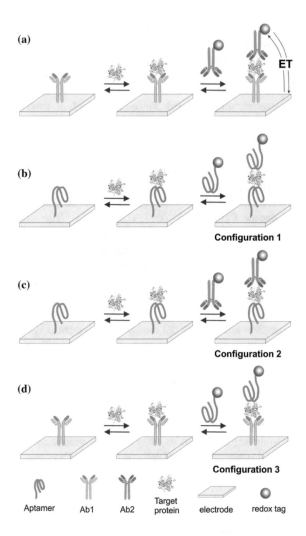

**Fig. 4.1** Schematic description of two classes of electrochemical sandwich assays for protein detection. Typically, the sandwich assay platform is composed of three elements: capture elements, target protein, and probe elements. Capture elements are immobilized on electrode surface, followed by target protein binding. This complex binds to a probe element pre-labeled with a redox tag, forming a sandwich structure. These capture and elements for protein detection are antibodies or aptamers. **a** Shown is antibody–antigen–antibody immunoassay. **b, c,** and **d** Shown are aptamer-based sandwich assays

magnetic beads (MBs) and nanoparticles, DNA-based hybridization chain reaction (HCR). In this chapter, we will focus on these amplification strategies for protein detection based on the aforementioned two classes of the electrochemical sandwich-type biosensors: immunoassay and aptamer-based assay.

## 4.2 Immunoassays

The antibody–antigen–antibody immunoassay requires the simultaneous binding or two antibodies with two different epitopes of the antigen, rendering them highly selective and specific even in complex clinic samples [11]. In order to achieve nanomolar or lower detection limits, this assay is commonly coupled with several amplification strategies, such as enzyme amplification, metal nanoparticles, carbon-based nanomaterials, MBs, DNA-based HCR. In the following section, we will review the unique feature and application of these main strategies individually, as well as some of the examples with a combination of two or three amplification approaches, such as an enzyme nanoparticles or a carbon nanotube–Au nanoparticles dual amplification.

### 4.2.1 Enzyme Amplification

Enzymes, like horseradish peroxidase (HRP), alkaline phosphatase (ALP), glucose oxidase (GOD), and lactate dehydrogenase, are commonly employed in biosensor platform for signal amplification [12]. In a typical configuration of enzyme-labeled electrochemical sandwich assay, the antigen binds to pre-immobilized antibody on the electrode substrate, followed by a second enzyme-labeled antibody binding event. The enzyme on second antibody provides a transducing signal as a quantification of target antigen. The electroactive products generated via an enzymatic oxidation or reduction of additional substrates provides electrochemical signals with a much higher signal-to-noise ratio. For example, Akanda et al. reported an ALP-based amplification strategy for cardiac troponin I detection [13]. Specifically, they immobilized the indium tin oxide electrode with the first antibody via a biotin–avidin interaction, which further formed a sandwich complex by binding to an antigen together with a second antibody pre-labeled by ALP (Fig. 4.2). This immunoreaction can be quantified by enzymatic reaction of appropriate substrates, further enhanced by a redox cycling when introducing a reducing agent, here, tris (2-carboxyethyl)-phosphine (TCEP).

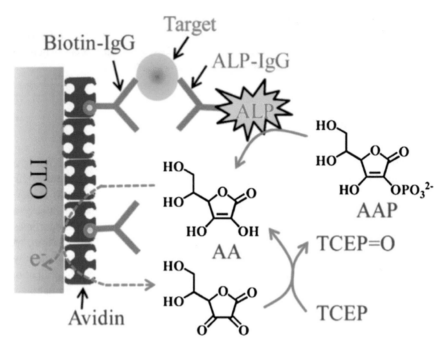

**Fig. 4.2** Schematic representation of an electrochemical immunosensor using the generation of L-ascorbic acid by alkaline phosphatase and the redox cycling of L-ascorbic acid by TCEP. (Reprinted with the permission from Ref. [13]. Copyright 2011 American Chemical Society)

HRP is another commonly used enzyme for electrochemical sandwich-type biosensors [14, 15]. Zhao et al. reported an electrochemical immunosensor designed for immunoglobulin G (IgG) detection using HRP-loaded $SiO_2$ silica-poly(acrylic acid) brushes ($SiO_2$-SPAABs), and the complex redox label exhibited higher catalytic performance using o-phenylenediamine (OPD) as a mediator (Fig. 4.3) [16]. The binding event generated a significant change in DPV current, supporting for ultrasensitive determination of target antigen. In contrast to conventional ELISA assay, this platform exhibited a sevenfold higher sensitivity.

## 4.2.2 Metal Nanoparticles Amplification

Signal amplification based on enzyme typically suffers from its instability. For instance, most enzyme immunoassays are susceptible to the changes in environmental conditions, for example, pH or temperature variations. Alternatively, enzyme-catalytic mimicking nanomaterials are more stable, more convenient, and cost-effective.

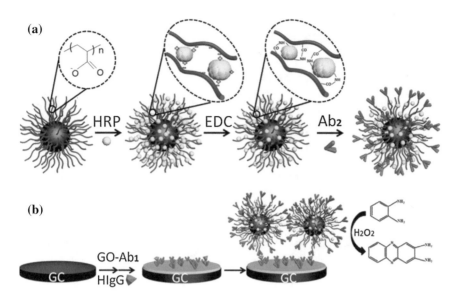

**Fig. 4.3** Schematic illustration of the electrochemical immunoassay protocol. **a** Covalent immobilization of HRP into $SiO_2$-SPAABs by chemical conjugation of antibody onto this nanobrush via N-hydroxysuccinimide and N-(3-dimethylaminopropyl)-N′-ethyl-carbodiimide hydrochloride (NHS/EDC) process, followed by secondary antibody immobilization via the same chemistry process. **b** $SiO_2$-SPAABs amplified sandwich-type immunosensing strategy and signal amplification method. (Reprinted with the permission from Ref. [16]. Copyright 2016 Elsevier)

Nanomaterials include noble metal nanoparticles, graphene, carbon nanotube, magnetic nanoparticles (MNPs), sharing a common feature of large surface area for attachment of a high surface concentration of capture antibodies, which could possibly provide electrocatalytic capabilities [17]. Nevertheless, each group of nanomaterial exhibits its own merits. For example, noble metal nanoparticles exhibit high stability, conductivity, biocompatibility, and catalytic activity, enabling them for the application of carrier or signal transducer, while carbon materials are mainly exploited as carrier and in the meanwhile can enhance the electron transfer efficiency. As a guideline, we will discuss their amplification strategy based on these different subgroups of nanomaterials in this and following sections.

Noble metal nanoparticles and nanomaterials are of great interests, as aforementioned, due to their high conductivity, high area surfaces, electrocatalytic capabilities, and strong electronic signal [18]. Depending on their functionality, metal nanomaterials can commonly be applied into immunoassay platform as two groups, one of which functions as a carrier for more effectively loading purpose and the second as signal transducer either directly applied as redox reporter or used as an electronic catalyst for the redox mediator [19, 20]. For example, You et al. demonstrated a signal amplification via employing Au nanoparticles as carriers for immobilizing with increased amounts of redox reporter (here, ferrocene) for glycoprotein [21]. Zhu's group has employed Ag-titanium hybrid nanoparticles

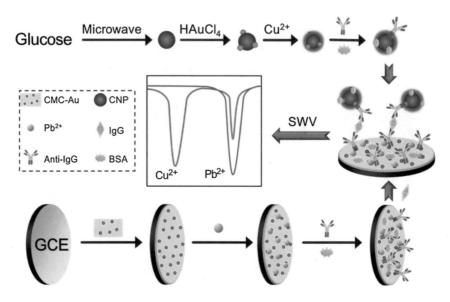

**Fig. 4.4** Schematic illustration of the function principle and the fabrication procedure of the proposed immunosensor. The stepwise fabrication procedures of secondary antibody-labeled Cu-coated Au nanoparticles (top) and capture antibody-immobilized substrates (bottom). The former employed Cu nanoparticles as the first redox label and the latter was modified by a second redox material, here, $Pb^{2+}$. The analyte can be detected simultaneously by the signals of substrate and probe, respectively (voltammograms, middle). The ratiometric detection strategy improved the detection sensitivity and accuracy. (Reprinted with the permission from Ref. [23]. Copyright 2016 Macmillan Publishers Ltd.)

(AgNP-TiP) as a redox label for human interleukin-6 detection [19]. Wang et al. developed a Pt porous nanoparticles-based braced structure with metal ion $Cd^{2+}$ and $Cu^{2+}$ as electrochemical signals. Using DPV technique, they achieved simultaneous detection of CEA and alpha-fetoprotein (AFP), exhibiting a dynamic range from 0.05 to 200 ng mL$^{-1}$ [22]. Tang et al. reported on a ratiometric approach for IgG detection via employing Cu@Au and Pb@Au nanoparticles as two redox materials with distinct redox potentials at 0.15 and $-0.45$ V by SV, respectively, with the former one placed on the electrode substrate and the latter as probe (Fig. 4.4) [23]. By measuring the ratio of peak currents of immunoprobe and substrate ($I_{probe}$/$I_{substrate}$), the proposed immunosensor for IgG detection exhibited a 2-orders of magnitude wider linear detection range (1 fg mL$^{-1}$–100 ng mL$^{-1}$) than other sandwich immunoassays.

Besides their roles directly as redox labels, metal nanoparticles have been frequently employed as electrocatalytic labels. For instance, gold (Au), platinum (Pt) nanoparticles, and bi- or multi-metallic nanostructures have been exploited for this purpose. Yang et al. employed Au nanoparticles to catalyze the redox reaction of the mediator, here *p*-nitrophenol, yielding a new oxidized species *p*-aminophenol. This new species can be catalytically recycled between its original structure and quinone form, achieving the goal of amplification [24]. Likewise, Pt nanoparticles have been exploited for sandwich immunoassay due to their good catalytic properties, e.g., oxygen reduction reaction, hydrogen evolution, and small organic mediator reduction [25–29]. Spain et al. explored a signal amplification based on electrocatalytic Pt nanoparticles for PSA detection. Specifically, the recombinant antibody was pre-labeled with Pt nanoparticles, and sandwiched antibody–antigen binding event promoted the catalytic reaction between Pt nanoparticles with peroxide, giving a measurable change in faradaic current associated with the concentration of PSA [30]. As it is true for the catalytic reaction for Au nanoparticles, Pt nanoparticles are able to catalyze $p$-nitrophenol reduction in the presence of $NaBH_4$ [27].

## 4.2.3 Carbon-Based Nanomaterial Amplification

Carbon-based nanomaterials have been exploited for biosensor platform because of their large surface area, their unique mechanical, electrical and optical properties [31]. We can divide them into three classes based on the dimensions of carbon nanomaterials: zero-dimensional materials, such as fullerenes and carbon quantum dots, one-dimensional (1D) carbon nanotube (CNT) (single or multi-walled carbon nanotubes (SWCNT or MWCNT)), and two-dimensional (2D) graphene. Carbon-based nanomaterials mainly serve as carrier for loading thousands of copies of labels or receptor antibodies, due to their more homogeneous, uniform, and large surface area compared to other nanoparticles, and as such, the immunoassay relying on their amplification typically involves immobilization of redox label materials, mostly nanoparticles. The use of CNTs and graphene has been vastly extended by tuning the physicochemical properties through modification of their surface, while the zero-dimensional carbon materials are much less explored in the voltammetric biosensing applications. For this reason, this chapter will mainly focus on the application of 1D and 2D carbon nanomaterials for amplifying signal in electro-chemical protein-detecting biosensors.

CNTs have been widely used as an ideal material for signal amplification, due to their low background noise, high tensile strength, high stability, low cost, and impressive electrical conductivity. The strong $\pi$–$\pi$ interactions between CNTs and the immobilized materials such as redox probes or receptors enable the robust attachment between them, while still maintaining CNTs' electrical conductivity for rapid electron transfer. Several strategies have been developed as employing carbon nanotube modification for immunoassay amplification, including redox label immobilization (strategy I), flat surface alignment (strategy II) [32–34], and forest alignment (strategy III) [35, 36] for electrode modifications. Wang et al. demonstrated for the first time the application of CNTs in loading numerous copies of redox labels for IgG detection (Fig. 4.5a) [37]. The numerous loading amount of ALP per CNT improved the detection sensitivity by 3-orders of magnitude in comparison with the single-ALP protocols. Similarly, glucose oxidase

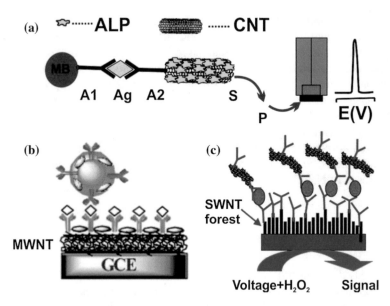

**Fig. 4.5** Three strategies using CNTs have been explored for signal amplification. **a** CNTs are employed as loading supports for redox tags, here ALP enzyme. Capture of the ALP-loaded CNT tags to the streptavidin-modified magnetic beads via an antibody–antigen–antibody interaction. Enzymatic reaction supported for electrochemical detection of the product of the enzymatic reaction at the CNT-modified glassy carbon electrode (Reprinted with the permission from Ref. [37]. Copyright 2011 American Chemical Society). **b** Flat alignment of MWCNTs serves as a loading support for capturing primary antibodies (Reprinted with the permission from Ref. [40]. Copyright 2015 The Royal Society of Chemistry). **c** Forest alignment of SWNT serves as loading support and electron transfer conductor (Reprinted with the permission from Ref. [41]. Copyright 2011 American Chemical Society)

(GOx) functionalized CNT as detection antibody label was employed for sensitive detection of IgG [38]. In another study, Ag nanoparticle–immobilized CNT (CNT/Ag NP) was used as redox label for simultaneous detections of multiple tumor markers, CEA, and α-fetoprotein, achieving acceptable precision and exhibiting detection limits of 0.093 and 0.061 pg mL$^{-1}$, respectively [39].

The second and third strategies are performed by aligning CNTs on the immobilizing surfaces either in a flat or forest configuration. In comparison with the flat alignment, the vertically aligned CNTs are more densely packed; in the meanwhile, they serve as molecular conductors, both of which effects could enhance electron transfer between the electrode surface and redox label. Qin et al. developed an electrochemical immunoassay for IgG detection via flatly aligning carbon nanotube as an immobilizing substrate (Fig. 4.5b) [40]. These modified electrodes were interrogated by LSV, showing a redox signaling of Au nanoparticles dissolution in the presence of HBr–Br$_2$ upon the advent of immunoreaction. In another study, Yu et al. exploited a SWNT forest strategy coupled with secondary antibody-nanotube bioconjugates for PSA detection in undiluted calf serum and human serum samples (Fig. 4.5c) [41], providing a detection limit of 4 pg mL$^{-1}$ in former, and a comparable accurate detection in the latter media.

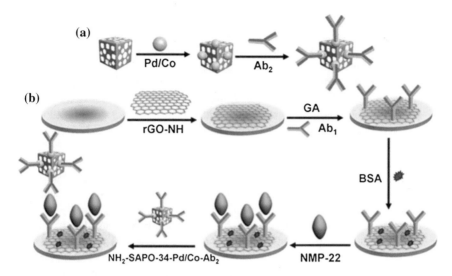

**Fig. 4.6** Schematic illustration of the fabrication procedure of the proposed graphene-based immunosensor. **a** The capture element, antibody-coated Pd/Co nanoparticles $NH_2$-SAPO-34-Pd/Co-$Ab_2$, were prepared by first absorbing Pd/Co nanoparticles onto molecular sieves support substrates and then immobilizing secondary antibodies onto these nanoparticles. Here, the nanoparticles were employed as redox labels. **b** The fabrication process of graphene-modified capture substrates was first by dropping rGO-NH solution onto electrode surface and immobilized by primary antibodies. After blocking with BSA, the platform recognized target protein and further bound the prefabricated secondary antibody-functionalized nanoparticles. (Reprinted with the permission from Ref. [42]. Copyright 2016 Macmillan Publishers Ltd.)

As it is true for CNTs, pristine graphene can be non-covalently functionalized through strong π–π interactions or covalently via classic chemistry strategies between its surface and the immobilized host, supporting the convenient and flexible functionalization. As such, they have been extensively employed into sandwich immunoassay as a carrier supporting substrate for either electrode surface modification immobilized with primary antibodies or redox tag loading for signal amplification. Wu et al. developed a novel sandwich-type electrochemical immunosensor for the detection of bladder cancer biomarker nuclear matrix protein-22 (NMP-22) (Fig. 4.6) [42]. They employed reduced graphene oxide-NH (rGO-NH) as sensing platform to immobilize primary antibody (Ab1), simultaneously promoting electrons transfer between the electrode surface and redox tag, here, molecular sieve supported Pd/Co nanoparticles (namely $NH_2$-SAPO-34-Pd/Co). The sensing platform was fabricated at three steps. First, rGO-NH solution was dropped and physically adsorbed onto electrode surfaces. Second, numerous copies of primary antibody Ab1 were immobilized onto rGO surface via chemical attachment between amine group of antibody and chitosan, followed by bovine serum albumin (BSA) immobilization for blocking purpose. Finally, the target antigen binds to primary antibody, and the immunoreaction happened following a second nanoparticle-labeled antibody binding. The redox signaling generated from nanoparticles Pd/Co labels reflects the concentration of target antigen. Such platform supports for sensitive detection of NMP-22 with a linear range from 0.001 to 20 ng/mL and a limit of detection (LOD) value of 0.33 pg/mL. Similar strategy was seen in many other studies, employing graphene as an immobilized substrate for primary antibodies [43–45]. Instead of using Pd/Co nanoparticles as redox labels, another study exploits gold nanocages (AuNCs) as conductive and loading platform for redox tags. Employing two district redox tags as signal transducers, such sensing platform have achieved simultaneous detection of autophagic proteins Belin-1 and LC3B-II with both linear ranges of 0.1–100 ng/mL [46].

Apart from its supporting substrate for primary antibodies, graphenes are also commonly used as loading substrate for redox tags [47–50]. Li et al. designed a sandwich electrochemical immunosensor for PSA detection using graphene and MWCNTs as redox label and capture antibody loading substrates, respectively [51]. The former was functionalized with Pd@Pt nanoparticle as redox label and secondary antibody for capturing purpose, and the latter MWCNTs was coated with AuNPs to increase load capacity of primary antibody. Both SWCNT and graphene materials, besides their roles in carriers, could also enhance the conductivity between redox label and electrode surface, thus enhancing the signal amplification. This platform supported sensitive detection of PSA target with a linear range from 10 fg/mL to 50 ng/mL and the LOD of 3.3 fg/mL.

### 4.2.4 Magnetic Beads and Magnetic Nanoparticles

MBs and MNPs have been widely used for the development of electrochemical immunosensors mainly for four reasons [52–58]: (1) their large surface-to-volume ratio and good biocompatibility, which allows them to serve as a loading scaffold for thousands of copies of antibodies or biomolecules of interests; (2) unique magnetic properties enabling rapid and easy separation of analytes under an external magnetic field, thus eliminating the washing steps and reducing the time of analysis for the complex real sample, such as blood serum or whole blood sample; (3) magnetic features enabling antibody-bearing MBs or MNPs directly load on electrode by a magnetic field, avoiding chemical linker modification onto the electrode; (4) due to their size effects, MNPs have an additional advantage, i.e., their intrinsic artificial enzyme mimetic activity allowing MNPs to catalyze the redox reaction of an appropriate mediator for signal amplification, as we discussed in Sect. 4.2.1.

Numerous examples of electrochemical sandwich-type immunoassay based on MBs or MNPs have been reported in the past decades. Typically, the platform of sandwich-type immunoassay is composed of an electrode which can be placed on a magnetic holding block, a primary antibody-modified MBs or MNPs that can be captured onto the electrode surface without any chemical linker, a redox transducer-labeled secondary antibody and the target antigen. Upon the binding of target antigen, electron transfer kinetics and/or the current between redox transducer and electrode surface will be changed, thus resulting in a measurable signal for target antigen detection using amperometric or other electrochemical techniques [59, 60]. For example, Torrente-Rodriguez et al. have developed a MBs scaffold on a screen-printed carbon electrode (SPCE) for fibroblast growth factor receptor 4 detection [60]. Briefly, they used MBs to capture the target protein and form sandwich architecture upon an enzyme-coupled secondary antibody binding. This platform exhibited detection limit of around 50 pg mL$^{-1}$. They also demonstrated the success of this platform to detect FGFR4 in cell lysates without apparent decrease performance. In another study, the same research group exploited a similar platform for p53 protein detection in different cell lysates, achieving a detection limit of 1.2 ng/mL [61]. Likewise, Lai et al. from another research group reported a similar strategy for IgG detection, replacing the enzyme signal transducer by thionine-doped mesoporous silica nanosphere (MSN)/polydopamine (PDA) nanocomposite. This sensing platform has achieved a wide linear detection range over four orders of magnitude and a low detection limit of 5.8 pg/mL [62].

MNPs are also widely reported as they can serve as loading support and magnetic separator (as for MBs), and in addition, they serve as electrocatalysts to amplify the transduction signals due to their smaller size, which have been discussed as for other metal nanoparticles (see Sect. 4.2.2) [63–65]. Wang et al. have developed an immunoassay for the determination of Nosema bombycis employing $Fe_3O_4$ MNPs, achieving a linear range from 0.001 to 100 ng mL$^{-1}$ [66]. Tang et al. combined the attributes of $Fe_3O_4$ nanoparticle and enzyme HRP, employing the former as primary antibody loading scaffold and the latter as redox signal transducer. Such advantageous platform promised a simultaneous detection of AFP and CEA in a single run [65].

### 4.2.5   Other Amplification Strategies

Besides these widely used amplification approaches summarized above, we will briefly discuss some other amplification strategies in the literature, such as DNA HCR, polymer, or dendrimer scaffold for loading substrates.

HCR strategy typically involves an initial DNA for in situ trigger many cycles of hybridization over dozens of copies of DNA, forming DNA concatemers, which are either pre-labeled with redox-active molecules or intercalated by redox indicators via pi–pi or electrostatic interactions (such as methylene blue, ferrocene, hexaammineruthenium (III) chloride, or other small redox molecules), thus providing signal amplification [67–71]. For example, Song et al. described an electrochemical immunosensor protocol with HCR signal amplification for sensitive detection of Epstein Barr virus nuclear antigen 1, employing doxorubicin hydrochloride as the intercalator reporter [72]. Using the same HCR strategy, Guo et al. have developed a sensing platform for simultaneous detection of AFP and PSA, employing DNA concatemers pre-labeled by ferrocene and methylene blue as their redox reporter, respectively. Under optimal conditions, their platform exhibited a linear range of 0.5 pg/mL–50 ng/mL for both targets [67].

Polymers or dendrimers, such as poly(amidoamine) (PAMAM) dendrimers, are of interests to serve as supporting substrates in electrochemical immunoassay mainly due to their multi-functionality and flexibility. For example, PAMAM is a monodisperse polymer with a globular shape and branched structure, containing adjustable number of copies of tertiary amine groups which can be used as the affinity support for metal nanoparticle assembly. Pei et al. have employed a PAMAM dendrimer-encapsulated CdS quantum dots as redox tags for IgG1 detection [73]. Sun has demonstrated a nanogold-encapsulated PAMAM electrochemical immunosensing platform for human carbohydrate antigen 19-9 detection [74].

## 4.3   Aptamer-Based Sandwich Assays

In the field of electrochemical biosensors, aptamers are promising alternatives to antibody for several reasons. First, aptamers are less expensive in its production process, and they are less susceptible to environmental conditions change than antibodies, such as variations in pH and temperatures. Due to their attributes, aptamers have been extensively explored in electrochemical sandwich assay for small molecules, protein biomarker detections. As an analogue to immunoassays, aptamer-based electrochemical sandwich assay can adopt three different configurations (Fig. 4.1b–d) depending on the functions of the aptamer on the platform as a capture, or redox label element, or both. In order to achieve the low detection limit of protein biomarker, effective amplification strategies are also required in the aptamer-based sandwich assay. Indeed, a variety of amplification strategies used in immunoassay have been likewise extensively explored for all the three configurations, such as enzyme catalyst, nanoparticles, carbon nanomaterials, DNA-based HCRs. In this section, we will review electrochemical aptamer-based sandwich assay based on these three configurations, on which the amplification strategies have been applied.

### 4.3.1   Configuration 1: Dual-Aptamer Binding Configuration

Typically, dual-aptamer binding sandwich assay requires a binding event of two aptamers to different regions of the same protein target (Configuration 1, Fig. 4.1b), one of which serves as capture probe and the other as reporter probe. The former is often assembled on the electrode via chemical adsorptions or magnetic field, while the latter is coupled with redox tags.

As is commonly seen in immunoassay, amplification strategies based on enzyme, metal nanoparticles, or carbon materials have been also extensively explored for aptasensor platform. Jing et al. employed GOD-immobilized graphene oxide as an enzyme redox label for thrombin detection. GOD could effectively catalyze the oxidation of glucose, further catalyzed by PdNPs and hemin/G-quadruplex, resulting in significant electrochemical signal amplification [75]. In another study, Salimi et al. reported a similar aptasensor for immunoglobulin E (IgE) detection, employing a pre-labeled capture aptamer with enzyme HRP as signal amplification. With this enzymatic amplification strategy, this aptasensor has achieved a low detection limit of 6 pM [76].

Very recently, Zhao et al. exploited cubic $Cu_2O$ nanocages loaded with Au nanoparticles as non-enzymatic electrocatalysts and redox reporter, in support to detecting thrombin in human serum sample. Interrogating with DPV, this aptasensor exhibited a sub-picomolar sensitivity and a linear detection range of 0.1 pmol/L–10 nmol/L [77]. Silver nanoparticles, as almost popular used as Au nanoparticles, have been mainly employed as signal transducers. For example, Song et al. introduced an aptasensor platform for PDGF detection using capture DNA-functionalized silver nanoparticles (AgNPs) as redox labels [78]. Likewise, Ocaña et al. adopted a sandwich assay employing Ag-coated Au nanoparticles as redox reporters for thrombin detection. In this study, instead of using traditional Au–S bonds, they employed streptavidin–biotin interaction to anchor the capture DNA probes onto Au nanoparticle surface [79].

Apart from enzyme and nanoparticles amplification, MBs or MNPs are seen in aptamer-based sandwich assay in favor of fast separation and avoiding sample pretreatment [80, 81]. Carbon materials, such as fullerene and CNTs, are used as loading scaffold and electron catalyst for signal enhancements [82, 83].

## 4.3.2   Configuration 2 and 3: Aptamer–Antibody Configuration

Provide that one cannot discover two aptamers that bind to the same target protein, the alternative solution is to replace one aptamer by an antibody (Configuration 2 and 3, Fig. 4.1c, d). These aptamer-antibody configurations are much less explored for protein detection, probably due to its complexity [84–86]. Again, these aptamer–antibody sensing platforms require signal amplification to achieve high sensitivity. Zamay et al. demonstrated the use of hydrophobic beads for signal enhancement in the aptasensor platform for lung cancer biomarker detection (Fig. 4.7) [85]. They employed screen-printed gold electrodes as the sensor platform for aptamer immobilization, which bound specifically to target protein, and coupled with the hydrophobic alkyl chain-modified silica-coated iron oxide MBs (IH-SAB-4C-8C hydrophobic beads) for SWV signal amplification. This sensor platform achieved a detection limit of 0.023 ng/mL, a 100-fold increase in

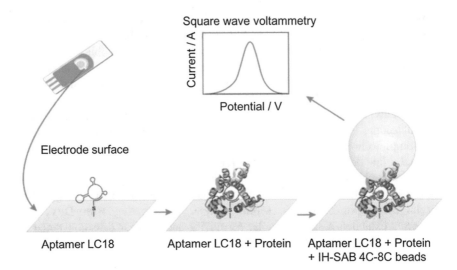

**Fig. 4.7** Scheme depicting an aptamer-based sensor for plasma protein detection. First, 5′-thiolatedDNA aptamer LC-18 is self-assembled onto a gold microelectrode (bottom left), followed by cancer-related blood plasma protein binding to the immobilized aptamer forming a complex (bottom middle). Finally, IH-SAB-4C-8C beads bind to the protein-enhancing SWV signal (bottom right). (Reprinted with the permission from Ref. [85]. Copyright 2016 Macmillan Publishers Ltd.)

comparison with those without beads. As such, this aptasensor promised a success in detection of target biomarkers in crude patient sample.

Ocaña et al. reported an aptamer-based sandwich assay for cytochrome c detection using multi-walled CNTs to immobilize the capture probes and Ag-coated Au nanoparticle for signal enhancement, exhibiting a detection limit of 12 pM [87]. In a similar assay, the same research group demonstrated the use of ALP as an enzyme label for lysozyme detection. By this enzyme-based signal amplification, they obtained a wide detection range and a detection limit of 4.3 fM, and this sensor platform was selective against other negative samples [88].

In contrast to the wide exploration of enzymatic and nanoparticles for aptamer–antibody sensing platform, MBs, and other amplification strategies, such as dendrimer, polymer-based, HCR, have been much less explored [86, 88, 89].

## 4.4 Conclusion

In this chapter, we reviewed the electrochemical sandwich assay for protein detection in the last decade, which could potentially provide an extensive and comprehensive road map for future guidelines. Two main classes of this assay, immunoassay and aptamer-based specific recognitions, appear as the most promising sensor platforms. The former employs antibody–antigen–antibody interactions, which is dominant due to this high specificity of simultaneous binding of dual antibodies to target antigen. In contrast, the aptamer-based sandwich assays are less explored probably due to their lower specificity and complexity in aptamer selections, despite their advantages over antibodies, such as their higher stability over environmental variations and lower cost for the production process.

These two classes of sensor platform require signal amplification in support to achieve sensitive detection of ultralow protein concentration (nanomolar or even lower range). These amplification strategies, including enzyme label amplification, nanoparticles, carbon nanotube, graphene, MBs, play an important role in signaling enhancement due to their unique electrical properties, large surface area, and electrocatalytic capabilities. These amplifiers can serve as loading support for immobilizing thousands of copies of redox labels or capture elements, due to their large surface area and easy functionality, such as nanoparticles, carbon nanotubes, and graphene. Likewise, they can serve as redox tags for signal transductions via promoting the catalysis of redox mediator and/or serving directly as redox species, such as enzymes, Ag nanoparticles, Pd, or Cd-containing metal nanoparticles.

We speculate that future directions of electrochemical sandwich assay would be translated into clinical  and point-of-care diagnostics. The main challenge remains as the direct detection of protein biomarker in real sample, for example, in patient plasma or whole blood sample, without any sample preparations. We believe that, the achievements of electrochemical sandwich-type platform summarized here would provide a comprehensive guideline for future explorations toward this goal.

# References

1. Smith RA, Cokkinides V, Brawley OW (2008) Cancer screening in the United States, 2008: a review of current American Cancer Society guidelines and cancer screening issues. CA Cancer J Clin 58:161–179
2. Dinarello CA (2007) Historical insights into cytokines. Eur J Immunol 37:S34–S45
3. Rong Q, Feng F, Ma Z (2016) Metal ions doped chitosan-poly(acrylic acid) nanospheres: Synthesis and their application in simultaneously electrochemical detection of four markers of pancreatic cancer. Biosens Bioelectron 75:148–154
4. Shen J, Li Y, Gu H, Xia F, Zuo X (2014) Recent development of sandwich assay based on the nanobiotechnologies for proteins, nucleic acids, small molecules, and ions. Chem Rev 114:7631–7677
5. Bonham AJ, Hsieh K, Ferguson BS, Vallee-Belisle A, Ricci F, Soh HT, Plaxco KW (2012) Quantification of transcription factor binding in cell extracts using an electrochemical, structure-switching biosensor. J Am Chem Soc 134:3346–3348
6. Guiseppi-Elie A, Lingerfelt L (2005) Impedimetric detection of DNA hybridization: towards near-patient DNA diagnostics. Top Curr Chem 260:161–186
7. Grieshaber D, MacKenzie R, Vörös J, Reimhult E (2008) Electrochemical biosensors—sensor principles and architectures. Sensors 8:1400–1458
8. Ren KW, Wu J, Yan F, Zhang Y, Ju HX (2015) Immunoreaction-triggered DNA assembly for one-step sensitive ratiometric electrochemical biosensing of protein biomarker. Biosens Bioelectron 66:345–349

9. Feng LN, Bian ZP, Peng J, Jiang F, Yang GH, Zhu YD, Yang D, Jiang LP, Zhu JJ (2012) Ultrasensitive multianalyte electrochemical immunoassay based on metal ion functionalized titanium phosphate nanospheres. Anal Chem 84:7810–7815

10. Solanki PR, Patel MK, Ali MA, Malhotra BD (2015) A chitosan modified nickel oxide platform for biosensing applications. J Mater Chem B 3:6698–6708

11. Zuo XL, Xiao Y, Plaxco KW (2009) High specificity, electrochemical sandwich assays based on single aptamer sequences and suitable for the direct detection of small-molecule targets in blood and other complex matrices. J Am Chem Soc 131:6944–6945

12. Yang H (2012) Enzyme-based ultrasensitive electrochemical biosensors. Curr Opin Chem Biol 16:422–428

13. Akanda MR, Aziz MA, Jo K, Tamilavan V, Hyun MH, Kim S, Yang H (2011) Optimization of phosphatase- and redox cycling-based immunosensors and its application to ultrasensitive detection of troponin I. Anal Chem 83:3926–3933

14. Du D, Wang LM, Shao YY, Wang J, Engelhard MH, Lin YH (2011) Functionalized graphene oxide as a nanocarrier in a multienzyme labeling amplification strategy for ultrasensitive electrochemical immunoassay of phosphorylated p53 (S392). Anal Chem 83:746–752

15. Khalilzadeh B, Shadjou N, Eskandani M, Charoudeh HN, Omidi Y, Rashidi M-R (2015) A reliable self-assembled peptide based electrochemical biosensor for detection of caspase 3 activity and apoptosis. RSC Adv 5:58316–58326

16. Zhao Y, Zheng YQ, Kong RM, Xia L, Qu FL (2016) Ultrasensitive electrochemical immunosensor based on horseradish peroxidase (HRP)-loaded silica-poly(acrylic acid) brushes for protein biomarker detection. Biosens Bioelectron 75:383–388

17. Rusling JF, Bishop GW, Doan NM, Papadimitrakopoulos F (2014) Nanomaterials and biomaterials in electrochemical arrays for protein detection. J Mater Chem B 2:12–30

18. Qin XL, Liu L, Xu AG, Wang LC, Tan YM, Chen C, Xie QJ (2016) Ultrasensitive immunoassay of proteins based on gold label/silver staining, galvanic replacement reaction enlargement, and in situ microliter-droplet anodic stripping voltammetry. J Phys Chem C 120:2855–2865

19. Peng J, Feng LN, Ren ZJ, Jiang LP, Zhu JJ (2011) Synthesis of silver nanoparticle-hollow titanium phosphate sphere hybrid as a label for ultrasensitive electrochemical detection of human interleukin-6. Small 7:2921–2928

20. Kong FY, Xu BY, Xu JJ, Chen HY (2013) Simultaneous electrochemical immunoassay using CdS/DNA and PbS/DNA nanochains as labels. Biosens Bioelectron 39:177–182

21. You M, Yang SA, Tang WX, Zhang F, He PG (2017) Ultrasensitive electrochemical detection of glycoprotein based on boronate affinity sandwich assay and signal amplification with functionalized $SiO_2$@Au nanocomposites. ACS Appl Mater Interfaces 9:13855–13864

22. Wang ZF, Liu N, Ma ZF (2014) Platinum porous nanoparticles hybrid with metal ions as probes for simultaneous detection of multiplex cancer biomarkers. Biosens Bioelectron 53:324–329

23. Tang ZX, Ma ZF (2016) Ratiometric ultrasensitive electrochemical immunosensor based on redox substrate and immunoprobe. Sci Rep 6:35440

24. Das J, Aziz MA, Yang H (2006) A Nanocatalyst-based assay for proteins: DNA-free ultrasensitive electrochemical detection using catalytic reduction of $p$-nitrophenol by gold-nanoparticle labels. J Am Chem Soc 128:16022–16023

25. Cui ZT, Wu D, Zhang Y, Ma HM, Li H, Du B, Wei Q, Ju HX (2014) Ultrasensitive electrochemical immunosensors for multiplexed determination using mesoporous platinum nanoparticles as nonenzymatic labels. Anal Chim Acta 807:44–50

26. Xu QN, Wang LS, Lei JP, Deng SY, Ju HX (2013) Platinum nanodendrite functionalized graphene nanosheets as a non-enzymatic label for electrochemical immunosensing. J Mater Chem B 1:5347–5352

27. Tang J, Zhou J, Li Q, Tang D, Chen G, Yang H (2013) In situ amplified electronic signal for determination of low-abundance proteins coupling with nanocatalyst-based redox cycling. Chem Commun 49:1530–1532

28. Fu XH, Huang R, Wang JX, Feng XR (2013) Platinum nanoflower-based catalysts for an enzyme-free electrochemical immunoassay of neuron-specific enolase. Anal Methods 5:3803–3806
29. Zhang J, Ting BP, Khan M, Pearce MC, Yang YY, Gao ZQ, Ying JY (2010) Pt nanoparticle label-mediated deposition of Pt catalyst for ultrasensitive electrochemical immunosensors. Biosens Bioelectron 26:418–423
30. Spain E, Gilgunn S, Sharma S, Adamson K, Carthy E, O'Kennedy R, Forster RJ (2016) Detection of prostate specific antigen based on electrocatalytic platinum nanoparticles conjugated to a recombinant scFv antibody. Biosens Bioelectron 77:759–766
31. Tiwari JN, Vij V, Kemp KC, Kim KS (2016) Engineered carbon-nanomaterial-based electrochemical sensors for biomolecules. ACS Nano 10:46–80
32. Lai GS, Zhang HL, Yong J, Yu AM (2013) In situ deposition of gold nanoparticles on polydopamine functionalized silica nanosphere for ultrasensitive nonenzymatic electrochemical immunoassay. Biosens Bioelectron 47:178–183
33. Yan ZQ, Ma HM, Fan DW, Hu LH, Pang XH, Gao J, Wei Q, Wang Q (2016) An ultrasensitive sandwich-type electrochemical immunosensor for carcino embryonie antigen based on supermolecular labeling strategy. J Electroanal Chem 781:289–295
34. Mazloum-Ardakani M, Hosseinzadeh L, Khoshroo A (2015) Ultrasensitive electrochemical immunosensor for detection of tumor necrosis factor-$\alpha$ based on functionalized MWCNT-gold nanoparticle/Ionic liquid nanocomposite. Electroanalysis 27:2518–2526
35. Chikkaveeraiah BV, Bhirde A, Malhotra R, Patel V, Gutkind JS, Rusling JF (2009) Single-wall carbon nanotube forest arrays for immunoelectrochemical measurement of four protein biomarkers for prostate cancer. Anal Chem 81:9129–9134
36. Malhotra R, Patel V, Vaqué JP, Gutkind JS, Rusling JF (2010) Ultrasensitive electrochemical immunosensor for oral cancer biomarker IL-6 using carbon nanotube forest electrodes and multilabel amplification. Anal Chem 82:3118–3123
37. Wang J, Liu GD, Jan MR (2004) Ultrasensitive electrical biosensing of proteins and DNA: carbon-nanotube derived amplification of the recognition and transduction events. J Am Chem Soc 126:3010–3011
38. Zhou M, Sun ZF, Shen CC, Li ZY, Zhang Y, Yang MH (2013) Application of hydrogel prepared from ferrocene functionalized amino acid in the design of novel electrochemical immunosensing platform. Biosens Bioelectron 49:243–248
39. Lai GS, Wu J, Ju HX, Yan F (2011) Streptavidin-functionalized silver-nanoparticle-enriched carbon nanotube tag for ultrasensitive multiplexed detection of tumor markers. Adv Funct Mater 21:2938–2943
40. Qin XL, Xu AG, Liu L, Deng WF, Chen C, Tan YM, Fu YC, Xie QJ, Yao SZ (2015) Ultrasensitive electrochemical immunoassay of proteins based on in situ duple amplification of gold nanoparticle biolabel signals. Chem Commun 51:8540–8543
41. Yu X, Munge B, Patel V, Jensen G, Bhirde A, Gong JD, Kim SN, Gillespie J, Gutkind JS, Papadimitrakopoulos F, Rusling JF (2006) Carbon nanotube amplification strategies for highly sensitive Immunodetection of cancer biomarkers. J Am Chem Soc 128:11199–11205
42. Wu D, Wang YG, Zhang Y, Ma HM, Yan T, Du B, Wei Q (2016) Sensitive electrochemical immunosensor for detection of nuclear matrix protein-22 based on $NH_2$-SAPO-34 supported Pd/Co nanoparticles. Sci Rep 6:24551
43. Zhou SW, Wang YY, Zhu JJ (2016) Simultaneous detection of tumor cell apoptosis regulators Bcl-2 and Bax through a dual-signal-marked electrochemical immunosensor. ACS Appl Mater Interfaces 8:7674–7682
44. Qin XL, Xu AG, Liu L, Sui YY, Li YL, Tan YM, Chen C, Xie QJ (2017) Selective staining of CdS on ZnO biolabel for ultrasensitive sandwich-type amperometric immunoassay of human heart-type fatty-acid-binding protein and immunoglobulin G. Biosens Bioelectron 91:321–327
45. Lin DJ, Wu J, Ju HX, Yan F (2014) Nanogold/mesoporous carbon foam-mediated silver enhancement for graphene-enhanced electrochemical immunosensing of carcinoembryonic antigen. Biosens Bioelectron 52:153–158

46. Wang GN, Li YK, Liu JL, Yuan YJ, Shen ZL, Mei XF (2017) Ultrasensitive multiplexed immunoassay of autophagic biomarkers based on Au/rGO and Au nanocages amplifying electrochemcial signal. Sci Rep 7:2442

47. Huang JL, Tian JN, Zhao YC, Zhao SL (2015) Ag/Au nanoparticles coated graphene electrochemical sensor for ultrasensitive analysis of carcinoembryonic antigen in clinical immunoassay. Sens Actuators B-Chem 206:570–576

48. Li L, Zhang LN, Yu JH, Ge SG, Song XR (2015) All-graphene composite materials for signal amplification toward ultrasensitive electrochemical immunosensing of tumor marker. Biosens Bioelectron 71:108–114

49. Wang D, Gan N, Zhang HR, Li TH, Qiao L, Cao YT, Su XR, Jiang S (2015) Simultaneous electrochemical immunoassay using graphene–Au grafted recombinant apoferritin-encoded metallic labels as signal tags and dual-template magnetic molecular imprinted polymer as capture probes. Biosens Bioelectron 65:78–82

50. Wen JL, Zhou SG, Yuan Y (2014) Graphene oxide as nanogold carrier for ultrasensitive electrochemical immunoassay of Shewanella oneidensis with silver enhancement strategy. Biosens Bioelectron 52:44–49

51. Li MD, Wang P, Li FY, Chu QY, Li YY, Dong YH (2017) An ultrasensitive sandwich-type electrochemical immunosensor based on the signal amplification strategy of mesoporous core-shell Pd@Pt nanoparticles/amino group functionalized graphene nanocomposite. Biosens Bioelectron 87:752–759

52. Luo Y, Asiri AM, Zhang X, Yang GH, Du D, Lin Y (2014) A magnetic electrochemical immunosensor for the detection of phosphorylated p53 based on enzyme functionalized carbon nanospheres with signal amplification. RSC Adv 4:54066–54071

53. Čadková M, Metelka R, Holubová L, Horák D, Dvořáková V, Bílková Z, Korecká L (2015) Magnetic beads-based electrochemical immunosensor for monitoring allergenic food proteins. Anal Biochem 484:4–8

54. Zarei H, Ghourchian H, Eskandari K, Zeinali M (2012) Magnetic nanocomposite of anti-human IgG/COOH-multiwalled carbon nanotubes/$Fe_3O_4$ as a platform for electrochemical immunoassay. Anal Biochem 421:446–453

55. de Souza Castilho M, Laube T, Yamanaka H, Alegret S, Pividori MI (2011) Magneto immunoassays for plasmodium falciparum Histidine-Rich Protein 2 related to malaria based on magnetic nanoparticles. Anal Chem 83:5570–5577

56. Yang ZH, Zhuo Y, Yuan R, Chai YQ (2016) Highly effective protein converting strategy for ultrasensitive electrochemical assay of Cystatin C. Anal Chem 88:5189–5196

57. Zhang HF, Ma LN, Li PL, Zheng JB (2016) A novel electrochemical immunosensor based on nonenzymatic Ag@Au-$Fe_3O_4$ nanoelectrocatalyst for protein biomarker detection. Biosens Bioelectron 85:343–350

58. Ho D, Sun XL, Sun SH (2011) Monodisperse magnetic nanoparticles for theranostic applications. Acc Chem Res 44:875–882

59. Ruiz-Valdepeñas Montiel V, Campuzano S, Conzuelo F, Torrente-Rodríguez RM, Gamella M, Reviejo AJ, Pingarrón JM (2015) Electrochemical magnetoimmunosensing platform for determination of the milk allergen β-lactoglobulin. Talanta 131:156–162

60. Torrente-Rodríguez RM, Ruiz-Valdepeñas Montiel V, Campuzano S, Pedrero M, Farchado M, Vargas E, Manuel de Villena FJ, Garranzo-Asensio M, Barderas R, Pingarrón JM (2017) Electrochemical sensor for rapid determination of fibroblast growth factor receptor 4 in raw cancer cell lysates. PLoS ONE 12:e0175056

61. Pedrero M, Manuel de Villena FJ, Muñoz-San Martín C, Campuzano S, Garranzo-Asensio M, Barderas R, Pingarrón JM (2016) Disposable amperometric immunosensor for the determination of human P53 Protein in cell lysates using magnetic micro-carriers. Biosensors 6:56

62. Lai GS, Zheng M, Hu WJ, Yu AM (2017) One-pot loading high-content thionine on polydopamine-functionalized mesoporous silica nanosphere for ultrasensitive electrochemical immunoassay. Biosens Bioelectron 95:15–20

63. Ge XX, Zhang AD, Lin YH, Du D (2016) Simultaneous immunoassay of phosphorylated proteins based on apoferritin templated metallic phosphates as voltammetrically distinguishable signal reporters. Biosens Bioelectron 80:201–207
64. Urbanova V, Magro M, Gedanken A, Baratella D, Vianello F, Zboril R (2014) Nanocrystalline iron oxides, composites, and related materials as a platform for electrochemical, magnetic, and chemical biosensors. Chem Mater 26:6653–6673
65. Tang J, Tang DP, Niessner R, Chen GN, Knopp D (2011) Magneto-controlled graphene immunosensing platform for simultaneous multiplexed electrochemical immunoassay using distinguishable signal tags. Anal Chem 83:5407–5414
66. Wang Q, Gan XX, Zang RH, Chai YQ, Yuan YL, Yuan R (2016) An amplified electrochemical proximity immunoassay for the total protein of Nosema bombycis based on the catalytic activity of $Fe_3O_4$NPs towards methylene blue. Biosens Bioelectron 81:382–387
67. Guo JJ, Wang JC, Zhao JQ, Guo ZL, Zhang YZ (2016) Ultrasensitive multiplexed immunoassay for tumor biomarkers based on DNA hybridization chain reaction amplifying signal. ACS Appl Mater Interfaces 8:6898–6904
68. Zhuo Y, Han J, Yu YQ, Chai YQ, Yuan R (2014) Signal amplification strategy with synergistic catalysis of hollow Pt nanochains and hemoglobin for electrochemical immunosensor. J Electrochem Soc 161:B26–B30
69. Ge YQ, Wu J, Ju HX, Wu S (2014) Ultrasensitive enzyme-free electrochemical immunosensor based on hybridization chain reaction triggered double strand DNA@Au nanoparticle tag. Talanta 120:218–223
70. Zhou J, Lai WQ, Zhuang JY, Tang J, Tang DP (2013) Nanogold-functionalized DNAzyme concatamers with redox-active intercalators for quadruple signal amplification of electrochemical immunoassay. ACS Appl Mater Interfaces 5:2773–2781
71. Zhang B, Liu BQ, Tang DP, Niessner R, Chen GN, Knopp D (2012) DNA-based hybridization chain reaction for amplified bioelectronic signal and ultrasensitive detection of proteins. Anal Chem 84:5392–5399
72. Song C, Xie GM, Wang L, Liu LZ, Tian G, Xiang H (2014) DNA-based hybridization chain reaction for an ultrasensitive cancer marker EBNA-1 electrochemical immunosensor. Biosens Bioelectron 58:68–74
73. Pei XM, Xu ZH, Zhang JY, Liu Z, Tian JN (2013) Sensitive electrochemical immunoassay of IgG1 based on poly(amido amine) dendrimer-encapsulated CdS quantum dots. RSC Adv 3:16410–16415
74. Sun AL (2015) Sensitive electrochemical immunoassay with signal enhancement based on nanogold-encapsulated poly(amidoamine) dendrimer-stimulated hydrogen evolution reaction. Analyst 140:7948–7954
75. Jing P, Yi HY, Xue SY, Chai YQ, Yuan R, Xu WJ (2015) A sensitive electrochemical aptasensor based on palladium nanoparticles decorated graphene–molybdenum disulfide flower-like nanocomposites and enzymatic signal amplification. Anal Chim Acta 853:234–241
76. Salimi A, Khezrian S, Hallaj R, Vaziry A (2014) Highly sensitive electrochemical aptasensor for immunoglobulin E detection based on sandwich assay using enzyme-linked aptamer. Anal Biochem 466:89–97
77. Zhao JM, Zheng T, Gao JX, Guo SJ, Zhou XX, Xu WJ (2017) A sub-picomolar assay for protein by using cubic $Cu_2O$ nanocages loaded with Au nanoparticles as robust redox probes and efficient non-enzymatic electrocatalysts. Analyst 142:794–799
78. Song W, Li H, Liang H, Qiang WB, Xu DK (2014) Disposable electrochemical aptasensor array by using in situ DNA hybridization inducing silver nanoparticles aggregate for signal amplification. Anal Chem 86:2775–2783
79. Ocaña C, del Valle M (2014) Signal amplification for thrombin impedimetric aptasensor: Sandwich protocol and use of gold-streptavidin nanoparticles. Biosens Bioelectron 54:408–414

80. Song W, Niu QQ, Qiang WB, Li H, Xu DK (2016) Enzyme-free electrochemical aptasensor by using silver nanoparticles aggregates coupling with carbon nanotube inducing signal amplification through electrodeposition. J Electroanal Chem 781:62–69

81. Wang Y, He X, Wang K, Ni X, Su J, Chen Z (2011) Electrochemical detection of thrombin based on aptamer and ferrocenylhexanethiol loaded silica nanocapsules. Biosens Bioelectron 26:3536–3541

82. Bai LJ, Chen YH, Bai Y, Chen YJ, Zhou J, Huang AL (2017) Fullerene-doped polyaniline as new redox nanoprobe and catalyst in electrochemical aptasensor for ultrasensitive detection of Mycobacterium tuberculosis MPT64 antigen in human serum. Biomaterials 133:11–19

83. Wang QQ, Zhou ZX, Zhai YL, Zhang LL, Hong W, Zhang ZQ, Dong SJ (2015) Label-free aptamer biosensor for thrombin detection based on functionalized graphene nanocomposites. Talanta 141:247–252

84. Taleat Z, Cristea C, Marrazza G, Mazloum-Ardakani M, Săndulescu R (2014) Electrochemical immunoassay based on aptamer–protein interaction and functionalized polymer for cancer biomarker detection. J Electroanal Chem 717:119–124

85. Zamay GS, Zamay TN, Kolovskii VA, Shabanov AV, Glazyrin YE, Veprintsev DV, Krat AV, Zamay SS, Kolovskaya OS, Gargaun A, Sokolov AE, Modestov AA, Artyukhov IP, Chesnokov NV, Petrova MM, Berezovski MV, Zamay AS (2016) Electrochemical aptasensor for lung cancer-related protein detection in crude blood plasma samples. Sci Rep 6:34350

86. Qureshi A, Gurbuz Y, Niazi JH (2015) Capacitive aptamer–antibody based sandwich assay for the detection of VEGF cancer biomarker in serum. Sens Actuators B-Chem 209:645–651

87. Ocaña C, Lukic S, del Valle M (2015) Aptamer-antibody sandwich assay for cytochrome c employing an MWCNT platform and electrochemical impedance. Microchim Acta 182:2045–2053

88. Ocana C, Hayat A, Mishra R, Vasilescu A, del Valle M, Marty J-L (2015) A novel electrochemical aptamer-antibody sandwich assay for lysozyme detection. Analyst 140:4148–4153

89. Zhang J, Yuan YL, BiXie S, Chai YQ, Yuan R (2014) Amplified amperometric aptasensor for selective detection of protein using catalase-functional DNA–PtNPs dendrimer as a synergetic signal amplification label. Biosens Bioelectron 60:224–230

# Chapter 5
# Sandwich Assays Based on SPR, SERS, GMR, QCM, Microcantilever, SAW, and RRS Techniques for Protein Detection

**Shenshan Zhan, Xiaoding Lou, Pei Zhou and Fan Xia**

**Abstract** Among the methods developed for protein sandwich assays, the strategies based on fluorescence, electrochemistry, and color change occupy the predominant portion. However, besides these three major types, there are some other techniques, such as use of surface plasmon resonance (SPR), surface-enhanced Raman scattering (SERS), giant magnetoresistive (GMR), quartz crystal microbalance (QCM), microcantilever, surface acoustic wave (SAW), and resonance Rayleigh scattering (RRS), which also play a very important role in the development of the sandwich assay for protein detection. Through integrating different recognition molecules, such as antibodies and aptamers, with conventional or new immerging sensing platforms, these assays exhibit excellent comparable sensitivities and specificities and attract extensive attention. Thus, in this chapter, some recent advances in these fields are summarized and concluding remarks on parts of which should be improved as well as outlook are outlined.

---

The original version of this chapter was revised: Foreword has been included and authors' affiliations have been updated. The erratum to this chapter is available at https://doi.org/10.1007/978-981-10-7835-4_13

---

S. Zhan · X. Lou · F. Xia
Hubei Key Laboratory of Bioinorganic Chemistry & Materia Medica,
School of Chemistry and Chemical Engineering, Huazhong University of Science and Technology, Wuhan 430074, People's Republic of China
e-mail: hbhgzss@hust.edu.cn

F. Xia
e-mail: xiafan@cug.edu.cn; xiafan@hust.edu.cn

X. Lou (✉) · F. Xia
Engineering Research Center of Nano-Geomaterials of Ministry of Education,
Faculty of Material Science and Chemistry, China University of Geosciences,
Wuhan 430074, People's Republic of China
e-mail: louxiaoding@cug.edu.cn; louxiaoding@hust.edu.cn

P. Zhou
School of Agriculture and Biology, Shanghai Jiao Tong University,
Shanghai 200240, People's Republic of China
e-mail: zhoupei@sjtu.edu.cn

© Springer Nature Singapore Pte Ltd. 2018
F. Xia et al. (eds.), *Biosensors Based on Sandwich Assays*,
https://doi.org/10.1007/978-981-10-7835-4_5

69

**Keywords** Protein detection · Sandwich assays · Surface plasmon resonance
Surface-enhanced Raman scattering · Giant magnetoresistive · Quartz crystal
microbalance · Microcantilever · Surface acoustic wave · Resonance Rayleigh
scattering

## 5.1  Introduction

By integrating the sandwich assays which on basis of the currently most utilized
recognition molecules in biochemical analysis, antibodies, and aptamers, with
multifarious sensing techniques including surface plasmon resonance (SPR),
surface-enhanced Raman scattering (SERS), giant magnetoresistive (GMR), quartz
crystal microbalance (QCM), microcantilever, surface acoustic wave (SAW), and
resonance Rayleigh scattering (RRS), various proteins of interest have been
specifically detected with ultrasensitivity. Depend on the employed techniques,
selected examples of these sandwich assays were classified into seven types in this
chapter. And the skeleton is as follows: The SPR-based sandwich assays were
discussed in accordance with the involved recognition molecules, the SERS-based
sandwich assays were illustrated according to the composition of the applied
nanostructures, the GMR-based sandwich assays were highlighted by listing representative reports of three different research groups, the QCM-based sandwich
assays were summarized in view of whether QCM has been used in parallel with
other techniques, the microcantilever-based sandwich assays were overviewed
according to the kind of the signal enhancement methods, and a few examples of
SAW/RRS-based sandwich assays were briefly introduced at last.

## 5.2  Sandwich Assays Based on SPR

Since the image of microscopic interfacial structure based on plasmon surface
polariton field was realized in 1988 due to the invention of surface plasmon
microscopy by Rothenhäuslar and Knoll [1], vast applications and fast increasing
research interests of the SPR sensors have been witnessed in various fields [2–6].
The working principle of a SPR sensor is based on an optical phenomenon in which
the interactive electromagnetic field at a dielectric/metal interface firstly excited the
collective coherent oscillations of free electrons in the conduction band of a metal,
creating the charge density oscillations which are known as surface plasmon
polaritons and forming an electric field which decays exponentially into its surrounding media with a penetration depth of hundreds of nanometers. As this
evanescent field is extremely sensitive toward the refractive index change of the
surrounding media, once the refractive index of the sensing medium changes the
characteristics of the incident light (e.g., angle, phase, wavelength), the beam for
SPR excitation will change accordingly (Fig. 5.1a) [6]. Recently, recognition
molecules including antibodies and aptamers have been applied to selectively
recognize proteins in sandwich assay-based SPR sensors. As these assays could be
divided into antibody–antibody sandwich format, aptamer–antibody sandwich

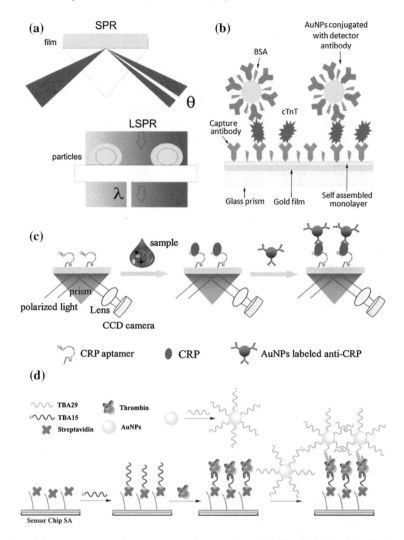

**Fig. 5.1** Schematic representations of **a** the sensing principle of the PSPR (above) and LSPR (below) systems. In the PSPR system, the SPR was excited on continuous metal thin films and could propagate along the surface up to hundreds of micrometers. While in the LSPR system, nonpropagating surface plasmon was excited on nanostructured metal surfaces (Reprinted with permission from Ref. [16]. Copyright 2016 Elsevier); **b** an AuNPs enhanced SPR sensor based on an antibody–antibody sandwich assay (Reprinted with permission from Ref. [10]. Copyright 2016 Elsevier); **c** a SPR sensor using aptamer–antibody sandwich assay and AuNPs conjugated with detector antibody for signal enhancement (Reprinted with permission from Ref. [17]. Copyright 2016 The Royal Society of Chemistry); **d** an AuNP-enhanced SPR sensor for thrombin detection with an aptamer–aptamer sandwich assay (Reprinted with permission from Ref. [18]. Copyright 2013 Elsevier)

format, and aptamer–aptamer sandwich format, according to the involved recognition molecules, those representative reports were discussed in accordance with this classification here in after.

Up to now, antibody–antibody sandwich assay is the prevailing types as most of the studied proteins have obtained their antibodies. Without using any signal amplification strategy, Altintas et al. constructed a SPR-based sandwich immunosensor for the cancer biomarker carcinoembryonic antigen (CEA) detection with a limit of detection (LOD) of 3 ng mL$^{-1}$ [7]. In order to enhance the SPR signals, nanomaterials including metallic nanoparticles [e.g., gold nanoparticles (AuNPs) and silver nanoparticles (AgNPs)], magnetic nanoparticles (MNPs), latex nanoparticles, carbon-based nanostructures, and liposome nanoparticles were often been exploited [6, 8]. Law et al. [9] established gold nanorod (AuNR)-enhanced biosensor for the detection of tumor necrosis factor alpha (TNF-α) antigen with a LOD of 0.03 pM by integrating both the immunoassay sensing technologies and nanoparticles into a SPR system. Solid AuNPs and hollow gold nanoparticles (HGNPs) have also been conjugated with special antibodies and employed as SPR signal amplifier in sandwich assays for ultrasensitive detection of cardiac troponin T (cTnT) (Fig. 5.1b) [10], human cardiac myoglobin (MYG) [11], infliximab [12], and cardiac troponin I(cTnI) [13] in serum samples at ng mL$^{-1}$ level. MNPs alone or coated with gold nanoshell also exhibited prominent SPR signal amplification effect in sandwich immunoassays for detections of β human chorionic gonadotropin [14] and human interleukin-17A [15], achieving LODs of 0.45 pM and 0.05 ng mL$^{-1}$, respectively.

As the potential alternatives to replace antibodies in drug development and medical diagnoses [19], aptamers have also been used to develop sandwich assay for protein detection. Chosen the immunoglobulin E (IgE) which has separate binding sites for aptamer and antibody as a model protein, Lee's group designed an aptamer–antibody sandwich assay platform. Detection was realized via the selective adsorption of untagged IgE proteins onto the surface which immobilized aptamer, followed by the specific adsorption of anti-IgE-coated AuNRs. Compared with the SPR measurements of the same sandwich format without using AuNRs, a remarkable $10^8$ enhancement which could measure IgE proteins at attomolar concentrations was achieved [20]. By substituting the AuNRs with gold chip, as well as changing the recognition molecules into aptamer and antibody which could specifically recognize the acknowledged biomarker for Alzheimer's disease, α-1 antitrypsin (AAT), the same group, introduced another aptamer–antibody sandwich assay for the SPR detection of AAT with a LOD of 10 fM [21]. Wu et al. presented an AuNP-enhanced SPR sensor for C-reactive protein (CRP) detection by utilizing an aptamer–antibody sandwich assay (Fig. 5.1c). The CRP-specific aptamers were firstly screened out applying a microfluidic chip and be immobilized on the Au surface to construct the SPR sensor. Then, CRP was introduced and captured through specific binding by these immobilized aptamers. Finally, anti-CRP-coated AuNPs were added to enhance the signal. As a result, CRP in diluted human serum could be detected selectively from 10 pM to 100 nM [17].

When designing sandwich assays for proteins such as thrombin which have more than one binding sites to their aptamers [22], it may consider completely taking the place of the antibodies with aptamers. Based on an aptamer/thrombin/aptamer–AuNP sandwich enhancement strategy, Bai et al. developed a SPR sensor for real-time detection of thrombin at subnanomolar level (Fig. 5.1d). In their protocol, one thiolated thrombin aptamer (TBA29) was fixed on AuNPs through Au–S bonding, and the other biotin-modified thrombin aptamer (TBA15) was linked onto streptavidin-preprocessed SPR gold film via streptavidin–biotin recognition. The presence of the target induced the formation of a double aptamer sandwich structure and resulted in significant enhancement of SPR signal. A good linear correlation in the thrombin concentration range of 0.1–75 nM and a LOD of 0.1 nM was obtained [18]. Through integrating the SPR imaging platform with near-infrared quantum dots, microwave-assisted surface chemistry, and aptamer technology, Vance et al. [23] designed an ultrasensitive sandwich-based assay for CRP detection at 5 fg mL$^{-1}$ level in spiked human serum. Nguyen et al. successfully screened out a few of aptamers for the whole avian influenza viruses H5Nx. Based on the isolated aptamers which could simultaneously recognize the different site of the same H5N1 whole virus, a highly specific and sensitive sandwich-format SPR-based sensor for the detection of H5Nx whole viruses was developed. The AuNPs were applied to enhance the signal, achieving a LOD of 200 50% egg infective dose per mL of inoculum (EID$_{50}$/mL) [24].

As in all the above-mentioned examples, the SPR was excited on continuous metal thin films and propagated along the dielectric/metal surface, these sensors could be classified into propagating SPR (PSPR) sensors [27]. There is a deep impression that in many literatures, unless pointed out particularly, the SPR sensor only refers to the PSPR sensor [28]. Besides these PSPR sensors, another main type of the SPR sensors is localized SPR (LSPR) sensors, in which nonpropagating surface plasmon was excited on nanostructured metal surfaces and localized SPR could be adjusted by their shape, size, and composition (Fig. 5.1a) [6]. Though compared with that of the PSPR sensor, spectral tunability of the LSPR sensor is better but sensitivity is orders of magnitude lower [29], which may explain why the number of the LSPR sensors is few, and reports on the application of LSPR in the sandwich assay-based protein detections also emerged in recent years. Haes et al. developed an optical biosensor to investigate the interaction between the amyloid-β-derived diffusible ligands (ADDLs) and the specific anti-ADDL antibodies based on LSPR spectroscopy and a sandwich format that consists of a primary antibody–antigen–secondary antibody conjugate (Fig. 5.2a) [25]. In their strategy, the target was captured by the primary antibody, while its molecular mass was augmented by the secondary antibody. Subsequently, a follow-up study in which the primary capture antibody was substituted by a thrombin binding aptamer (TBA) was reported (Fig. 5.2b) [26]. As the aptamer strand was usually much shorter than an antibody, the thrombin could be pulled to the nanoparticle surface more closely, causing an enhanced plasmonic peak shift. The LODs of this work without and with antibody enhancement were 18.3 and 1.6 pM, respectively, exhibiting a magnitude of improvement more than one order on the detection limit.

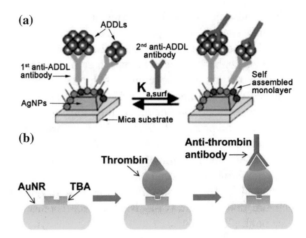

**Fig. 5.2** Schematic representations of LSPR sensors based on **a** antibody–antibody sandwich assay. Surface-confined AgNPs were firstly synthesized on mica substrates. Next, the first anti-ADDL antibody was covalently attached to the AgNPs which have been passivated for nonspecific binding and activated for the attachment of antibody by the self-assembled monolayer. After incubated in varying concentrations of ADDLs, the samples were incubated in the second anti-ADDL antibody solution to enhance the LSPR shift response of the ADDLs (Reprinted with permission from Ref. [25]. Copyright 2005 American Chemical Society); **b** antibody–aptamer sandwich assay. TBA-modified AuNRs were incubated with thrombin to trigger thrombin bind to the AuNRs. The thrombin binding event generated a LSPR shift, which was amplified by the incubation of the thrombin-bound AuNRs in a solution containing anti-thrombin antibodies. The binding of anti-thrombin to thrombin caused further local refractive index changes at the AuNR surface, generating secondary redshift of LSPR peak (Reprinted with permission from Ref. [26]. Copyright 2012 Elsevier)

## 5.3  Sandwich Assays Based on SERS

SERS-based sensing platforms have become one of the most promising approaches to detect target analytes including organic compounds, metal ions, nucleic acids, proteins, viruses, bacteria, and cancer cells due to their narrow line widths, large dynamic range, nondestructive testing capabilities, remarkable multiplexing ability, and excellent sensitivity/specificity [30, 31]. Briefly, SERS refers to the event in which the targets are in the vicinity of a nanostructure surface, dramatically enhancing ($10^2$–$10^{14}$) the Raman scattering, i.e., the inelastic scattering of a photon [32]. As the prerequisite in producing SERS event, the nanostructures employed are usually composed of metallic substrates such as single composition-based nano-materials (e.g., gold, silver, and copper) and composite-metal nanomaterials (e.g., alloy and magnetic nanostructures) [30]. Thus, in the following, some recently reported SERS-based sandwich assays for protein detection were illustrated according to the composition of the nanostructures.

Among the single composition-based nanomaterials, noble metal nanoparticles including AuNPs and AgNPs have found wide application in SERS because of their strong absorption of electromagnetic waves in the visible region, highly stable dispersions, biocompatibility, and chemical inertness [33, 34]. Wang et al. reported a single-step homogeneous SERS immunoassay for the detection of multiple target proteins based on controlled assembly of SERS nanoparticles (Fig. 5.3a) [35]. Characteristic of this method lies in that the SERS substrates (AuNRs or AuNPs) were codecorated by half-fragments of specifically orientated immobilized antibody, passivating proteins and nonfluorescent Raman-active dyes. As the construction of SERS nanoparticles resulted from the sandwiched antibody–antigen interactions, the interparticle distance was dramatically decreased and multicolor Raman fingerprint coding with high signal-to-noise ratio was provided. This platform has been applied for multiplexed quantification of interleukin-2, interferon gamma (IFN-$\gamma$), and TNF-$\alpha$, representing a potential tool for point-of-care clinical diagnostics and early detection of disease markers [36].

As another common substrate in SERS, AgNPs have been inevitably compared with AuNPs. Previous studies revealed that though the SERS spectra of the silver substrates are less stable than those of gold ones, the average SERS enhancement ability of AgNPs is much stronger than that of AuNPs [32, 40], and thus, the AgNPs have been used more widely. By functionalizing SERS-active AgNPs and polystyrene microspheres (PS) with monoclonal and polyclonal antibodies of the target antigens, respectively, Hwang et al. depicted the first optoelectrofluidic immunoassay platform for fast detection of alpha-fetoprotein (AFP). Through measuring the amount of probe nanoparticles on the AgNPs–antibody/antigen/ antibody–PS immunocomplexes, AFP in a sample droplet of $\sim 500$ nL could be detected within 5 min with a LOD of $\sim 0.1$ ng mL$^{-1}$ [41]. For the purpose of obtaining higher Raman signal, multistage signal amplification strategies such as the additional SERS effect of aggregated AgNPs and the high enzymatic activity of catalase have been applied to establish a aggregated AgNP-based SERS–ELISA method. With this method, ultralow levels of prostate-specific antigen (PSA) ($10^{-9}$ ng mL$^{-1}$) in whole serum could be detected [42]. High sensitivities have been achieved by these two methods; however, the most remarkable advantage of the SERS technique, i.e., multiplexing ability [43], has not been exploited. In view of this, Bazan and co-workers designed the SERS "antitags" which were capable of simultaneous detection of CRP, MYG, and human $\alpha$-thrombin with sensitivities of 100 pM by employing sandwich immunoassay and three different Raman reporters [44].

Aptamers have also been utilized in SERS-based sandwich protein detection as the widely accepted substitution of antibody. Before reporting the "antitags," Bazan's group designed the "aptatags" which consist of aptamer-modified AgNPs held together by organic dithiol molecule. A heterogeneous method for protein recognition was established by taking advantage of the sensing abilities of the aptamers and the Raman signal amplification effect of the AgNPs. For the particular example of thrombin, a LOD of 100 pM was attained [45]. Zengin et al. proposed a SERS aptasensor for the determination of ricin B toxin based on 4,4'-bipyridyl-labeled AgNPs and ricin B aptamer. The signal was produced by the sandwich assay took place between the ricin B toxin and the aptamers tagged on the

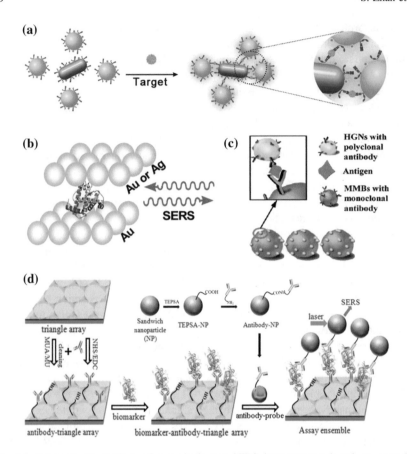

**Fig. 5.3** Schematic illustrations of **a** a single-step SERS immunoassay based on controlled assembly of SERS nanoparticles for the detection of multiple target proteins including interleukin-2, IFN-γ, and TNF-α (Reprinted with permission from Ref. [35]. Copyright 2013 American Chemical Society); **b** a SERS-based protein identification on basis of the interactions between proteins and AuNPs (Reprinted with permission from Ref. [37]. Copyright 2009 American Chemical Society); **c** formation of sandwich immunocomplex between HGNs and MBs. Polyclonal and monoclonal antibodies of the target proteins were conjugated onto the surfaces of HGNs and MBs, respectively, and sandwich immunocomplexes would form when mixing HGNs and MBs with target protein (Reprinted with permission from Ref. [38]. Copyright 2010 American Chemical Society); **d** principle of a SERS immunosensor for VEGF detection. After the capture antibody and the detection antibody were conjugated to the nanoarray chip and the sandwich nanoparticle by the carbodiimide chemistry, respectively, the capture antibody-modified Au chip was immersed into a solution containing the biomarker. During the incubation, the biomarker bound to the capture antibody-modified Au triangle nanoarray chip. Next, the biomarker antibody Au nanoarray chip was incubated in a solution containing the Au star@MGITC@SiO$_2$ nanoparticle conjugated with the detection antibody. Since the antigen in the present work had at least two binding sites, it can bind to both the detection antibody and the capture antibody, leading to the formation of the nanoparticle/biomarker/chip sandwich architecture. Finally, the chips were illuminated with the laser, and the SERS signal from the MGITC was recorded. (Reprinted with permission from Ref. [39]. Copyright 2013 American Chemical Society)

AgNPs as well as that immobilized on hybrid silicon substrate, and enhanced by silver deposition performed on the sandwich complex. As a result, linear relationship in the ricin B concentration range of 1.0 fM to 50 pM and a LOD of 0.32 fM was acquired [46].

Besides these aforesaid examples which adopted antibodies/aptamers as the SERS probes, there are also cases in which antibodies/aptamers have been replaced by small molecules and even been put aside. Relying on the sandwich formation between boronate-affinity SERS probes, targets, and boronate-affinity molecularly imprinted polymers (MIPs), Liu's group demonstrated a method for the sensitive and specific detection of trace glycoproteins. The MIP guaranteed the specificity, while the SERS detection provided the sensitivity with a LOD of 1 ng mL$^{-1}$ [47]. Han et al. developed a SERS-based protein identification strategy on the basis of the strong interactions between metal nanoparticles and proteins (Fig. 5.3b) [37]. Results showed that the SERS signals of antihuman IgG from the Ag/Au sandwich architectures was 7 times stronger than those from the Au/Au sandwiches. Based on the phenomenon that proteins could be sandwiched between the cetyltrimethylammonium bromide-coated positively charged AgNPs and citrate-reduced negatively charged AgNPs, brought about increased electromagnetic field and rising SERS signal, Yang et al. [48] presented two label-free fiber SERS sensors for cytochrome c and lysozyme detections in aqueous solution with LODs of 0.2 μg mL$^{-1}$, respectively. High sensitivities were claimed by these two reports which applied no SERS probes; nevertheless, specificities of them remain vaguely.

Magnetic nanostructures are ideal candidates for SERS-based sensing because of their novel features owing to the strong coupling between different components compared with single-component nanomaterials, as well as the presence of magnetic field which can efficiently induce electromagnetic enhancement hot spots [30, 49]. Choo's group reported several SERS-based immunoassay techniques, using magnetic beads (MBs) and hollow gold nanospheres (HGNs) or intact AuNPs. In the HGNs/SERS-based immunoassays they developed, the surfaces of HGNs and MBs were conjugated with polyclonal and monoclonal antibodies of the target proteins, respectively. Sandwich immunocomplexes would form when simply mixing the HGNs and MBs with target protein (Fig. 5.3c). Bar magnet or solenoids were applied to trap these immunocomplexes, followed by removing those nonspecific binding HGNs. By measuring the SERS signal around 1615 cm$^{-1}$, both of the IgG and CEA could be detected in 1–10 ng mL$^{-1}$ [38, 50]. Replacing the HGNs with intact AuNPs, the same group established another SERS-based immunoassay for the determination of free-to-total PSA ratio, of which the feasibility was verified by testing 30 clinical samples and compared the results with those measured by electrochemiluminescence instrument [51]. Besides spherical gold nanospheres, core–shell-structured nanocomposite and flowerlike AuNPs coupled to MBs also render them useful for biomarker detection due to the large surface area and abundant hot spots for SERS [52, 53].

Construction of noble metal alloy complex nanostructures has often been adopted when designing SERS-based assays as it could reasonably improve the performance of the noble metal [54]. Take Ag for example; it is the best plasmonic enhancement material among all the noble metals, but it gets oxidized easily in air, and thus, pure Ag nanostructures cannot exhibit satisfactory optical properties. Fortunately, Au possesses good biocompatibility, and it can be applied to hinder the oxidation of Ag while improving its stability [30]. In the past five years, several types of Ag/Au-based alloy nanostructures have been used in SERS-based sandwich assays for protein detection. Kong et al. exploited a simple technique for the synthesis of silica-encapsulated Ag (Ag@SiO$_2$) SERS tags which exhibited prominent stability resistance to long-term storage and high-concentration salts, depending on which assays of MYG conducted in two modes, both of which were based on the MYG-mediated complex formation between the iminodiacetic acids (IDAs) or IDA and glutaraldehyde, were achieved [55]. Hu et al. utilized the aptamers (Apt1 and Apt2) tagged core-shell nanoparticle architecture-silica coated silver (Ag@Si) NPs as the SERS enhancement substrate, and coomassie brilliant blue (R-250) as the Raman reporter and protein labeling, proposed a strategy for probe prion protein (PrP) detection. By monitoring the dramatically enhanced Raman signal aroused from the PrP/R-250-induced Ag@Si-PrP/R-250-Ag@Si conjugates, a linear equation was obtained in the range of 3–12 nM PrP [56]. Being adopted as highly sensitive SERS probes, antibody-labeled gold@silver core–shell nanorods (Au@Ag NRs) have been combined with microfluidic platform that immobilized with different antibodies to carry out multiplex immunoassays on a SERS-assisted 3D barcode chip. Through SERS scanning in each unit of the sandwich immunoassay, a 3D barcode containing the spectroscopic and spatial information with a supersensitivity down to 10 fgmL$^{-1}$ could be acquired [57]. Based on the Au nanostar@malachite green isothiocyanate (MGITC)@SiO$_2$ and Au triangle arrays, Li et al. constructed a SERS immunosensor for vascular endothelial growth factor (VEGF) detection in human blood plasma of patients (Fig. 5.3d). The capture antibodies were anchored to Au triangles on the substrate, and the detection antibodies were attached to Au nanostar SERS tags. Upon the target antigen sandwiched between the capture antibody and the detection antibody, large field enhancements occurred, bringing about an enhanced LOD of 7 fg mL$^{-1}$ [39].

## 5.4 Sandwich Assays Based on GMR

Since Fert [58] and Gruenberg [59] firstly discovered the GMR effect in 1988, and both were granted with the Nobel Prize of Physics for their contribution in 2007 [60], technologies based on GMR have witnessed a rapid development due to their extensive application including fluid control, nondestructive testing, electrical current measurement, and bioanalyte detection [61]. Among these applications, the GMR-based sandwich assay is one of the noteworthy representatives which could detect the stray magnetic fields resulted from the magnetic labels immobilized on the sensor surface owing to the formation of sandwich immunocomplex (Fig. 5.4a) [62, 63]. Different from traditional colorimetric or fluorescent sandwich assays, which suffer from interference because of the autofluorescence and inherent opacity

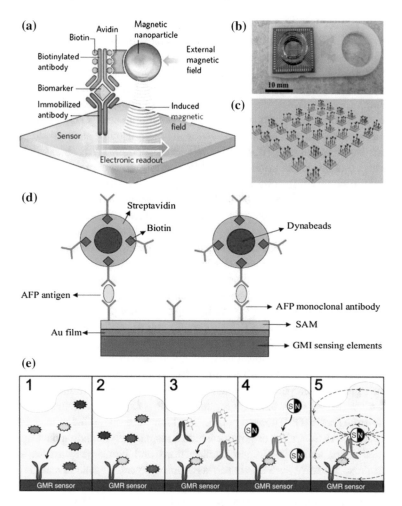

**Fig. 5.4** Schematic representations of **a** the mechanism of a GMR sensor based on sandwich assay (Reprinted with permission from Ref. [62]. Copyright 2009 Nature Publishing Group); **b** one reaction-well attached chip on the chip holder and **c** multiplex protein array on one chip (Reprinted with permission from Ref. [65]. Copyright 2015 Elsevier); **d** a complex immunoassay on Au film of the GMI biosensor for AFP detection (Reprinted with permission from Ref. [66]. Copyright 2014 Elsevier); **e** the sensing process of a sandwich assay-based GMR sensor. (1) capture antibodies were immobilized on the sensor surface; (2) nontarget antigens were washed away; (3) biotinylated antigen-specific detection antibodies were bonded in a sandwich structure, and the nonspecific ones were washed away; (4) streptavidin-tagged magnetic nanoparticles were added to the solution and bonded to the biotinylated detection antibodies; (5) magnetic fields from the nanoparticles were measured in the presence of a external magnetic field (Reprinted with permission from Ref. [67]. Copyright 2009 Nature Publishing Group)

of the circumambient matrices when detecting opaque biological samples, the sensitivity of GMR sensor could be enhanced as most of the surrounding biological matrix is intrinsic nonmagnetic [64].

Different research groups all over the world have reported a considerable number of GMR sandwich assays for protein detection. For example, the group of (Jian-Ping) Wang from the University of Minnesota established several high-magnetic-moment FeCo nanoparticles (12.8 nm) and GMR sensor-based detecting systems for the separate detection of biomolecules such as streptavidin [68], human interleukin-6 (IL-6) [68, 69], and endoglin [70]. Sandwich-based approaches that adopt the principle of ELISA with high specificity and sensitivity were used throughout. Results showed that signals from streptavidin as few as 600 copies, IL-6 as low as $2.08 \times 10^6$ molecules, and endoglin as low as 83 fM could be detected. Though great potential in helping to make rational treatment plan in disease development was shown by these detecting systems [71, 72], the quantitative analysis of a single biomarker is usually not qualified to identify the disease at an early stage. Thus, Wang's group developed an immune-biosensing system which was capable of real-time, multiplexed readout, working independently and be monitored simultaneously, based on a GMR chip (16 mm × 16 mm) that consisted of 64 nanosized sensor array (Fig. 5.4b, c) [65]. Three protein biomarkers (suppression of tumorigenicity 2 (ST2), proprotein convertase subtilisin/kexin type 9, and pregnancy-associated plasma protein A) had been analyzed on the basis of sandwich assay format, and four orders of magnitude of dynamic range for these three antigens with a LOD of 40 pg mL$^{-1}$ for ST2 had been attained.

Zhou's group developed two Dynabeads-labeled magnetic immunoassays for the detection of AFP (Fig. 5.4d) [66] and CRP [73] by utilizing two giant magneto-impedance (GMI)-based platforms which were integrated with the Cr/Cu/NiFe/Cu/NiFe/Al$_2$O$_3$/Cr/Au or gold substrates, respectively. Sandwich assays were performed using antibody–antigen combinations and biotin–streptavidin interactions on those substrate surface via self-assembled layers. Results revealed that the presence of the targets on the platform improved the GMI effect owing to the induced magnetic dipole of superparamagnetic beads. Both of these two assays have linear detection ranges in 1–10 ng mL$^{-1}$.

As for multiplexed detection, the pioneering approaches reported by (Shan X.) Wang's group from the Stanford University really represent the important progress toward practical application including basic science and clinical research. Sandwich assays were employed throughout, of which the target was sandwiched between one antibody that immobilized on the sensor surface and the other that labeled with a superparamagnetic nanoparticle through streptavidin–biotin interaction. The nanoparticle was magnetized using an external magnetic field, and the quantity of the target can be determined in the form of electronic readout which was proportional to the extent of nanoparticle binding (Fig. 5.4a) [62]. Based on this strategy, multiplex detections of potential cancer markers (eotaxin, lactoferrin, CEA, TNF-α, interleukin-10, granulocyte colony-stimulating factor (G-CSF), IFN-γ, and interleukin-1-alpha) at subpicomolar concentration levels [74], potential tumor markers (CEA, TNF-α, G-CSF, eotaxin, survivin, lactoferrin, VEGF, and epithelial cell adhesion molecule) down to the attomolar level (Fig. 5.4e) [67], mycotoxins (HT-2, zearalenone and aflatoxins B1) with LODs of 50 pg mL$^{-1}$ [75], food allergens (Gliadin, Ara h 1 and Ara h 2) at ng mL$^{-1}$ levels [76], and human IgG and IgM with sub-nM detection limits [77] were successfully achieved.

## 5.5 Sandwich Assays Based on QCM

QCM is a mass sensing platform making use of the piezoelectric properties of quartz crystals [78]. It can provide quantitative and qualitative information about molecular interactions by translating mass changes at the crystal sensor surface which immobilized the probe into detectable changes in the resonant frequency of the quartz crystal [79]. Due to its numerous advantages such as simple, label-free, cost-effective and high resolution, the QCM has been utilized as valuable tools for an extensive range of applications including material science and biosensing [80].

Henne et al. [81] described a QCM sensor for the determination of folate-binding protein (FBP) with a LOD of 30 nM, which was further improved to 50 pM by using an anti-FBP antibody and protein A-coated AuNP sandwich assay. Though AuNPs have been frequently used in signal amplification [82], the amplification mechanism of them has been seldom discussed. Chen et al. developed an aptamer/protein/aptamer–AuNP sandwich QCM biosensor for sensitive detection of human α-thrombin. Besides achieving a low LOD of 0.1 nM, the amplification effect of the aptamer–AuNPs on QCM as both mass and viscoelasticity enhancers was also demonstrated for the first time [83]. Other signal amplification strategies without using AuNPs have also been established. Based on a wireless electrode-less QCM with a fundamental resonance frequency of 182 MHz, Ogi et al. developed a mass sensitivity amplification method-coupled sandwich assay for CRP detection (Fig. 5.5a). The first CRP antibody was immobilized on the quartz surfaces non-specifically, the biotin-modified second anti-CRP antibody was weighted by streptavidin for sensitivity improvement, and the mass-amplified sandwich assay was triggered by injection of the CRP solution, with a LOD of 0.1 ng mL$^{-1}$ obtained [84].

QCM has been applied in parallel with numerous optical, microscopic, and spectroscopic techniques [85], and combination of QCM with these techniques can largely improve the precision with parallel measurements and acquire complementary details [86, 87]. Among those combined strategies, integration of SPR and QCM is one of the current trends [78]. Usually, strategies of this type share one similarity; i.e., AuNPs were applied to enhance the signal, in which the amplification of the SPR signals was realized by enlarging the refractive index of the surface, whereas the QCM signal was enhanced because of the mass increment brought by AuNPs. Uludag et al. developed an immunosensor to detect the total prostate-specific antigen (tPSA) in human serum samples using QCM and SPR platforms (Fig. 5.5b). Based on a sandwich assay that employing antibody-modified AuNPs, tPSA in 75% human serum at the concentrations of

**Fig. 5.5** Schematic illustrations of **a** the sandwich assay applying the mass-amplified second anti-CRP antibody. a, b, and ab denote the mass-amplified second antibody, the CRP adsorbed on the first antibody, and their complex, respectively (Reprinted with permission from Ref. [84]. Copyright 2011 Elsevier); **b** the PSA sandwich assay utilizing PSA antibody-modified AuNPs (Reprinted with permission from Ref. [88]. Copyright 2012 American Chemical Society)

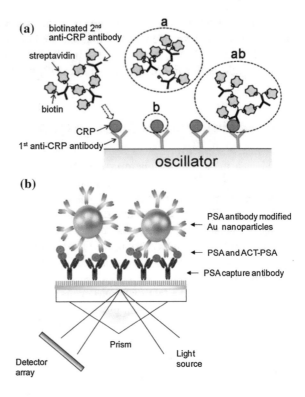

2.3 ngmL$^{-1}$ (AuNPs, 20 nm) and 0.29 ngmL$^{-1}$ (AuNPs, 40 nm) could be determined [88]. In combination with sandwich recognition mechanism, biobarcoded AuNP enhancement, and aptamer-based rolling circle amplification, He et al. proposed an SPR and QCM sensing platforms for ultrasensitive detection of human a-thrombin with a LOD low to 0.78 aM [78].

## 5.6  Sandwich Assays Based on Microcantilever

Microcantilevers have attracted wide attentions in the development of a great variety of miniaturized chemical, physical, and biological sensors owing to their advantages including low cost, quick response, high sensitivity, direct detection, and array capability [89]. The transduction mechanism of a microcantilever sensor is that intermolecular forces produced by molecular adsorption cause the mechanical bending and change the deflection and resonance properties of the cantilever (Fig. 5.6a) [90].

Etayash et al. presented a multiplexed approach that uses cantilever arrays for simultaneous detection of Glypican-1 (GPC1), EGFR, CD24, and CD63. A best performance with a LOD of 1 ng mL$^{-1}$ was attained when detecting GPC1, which

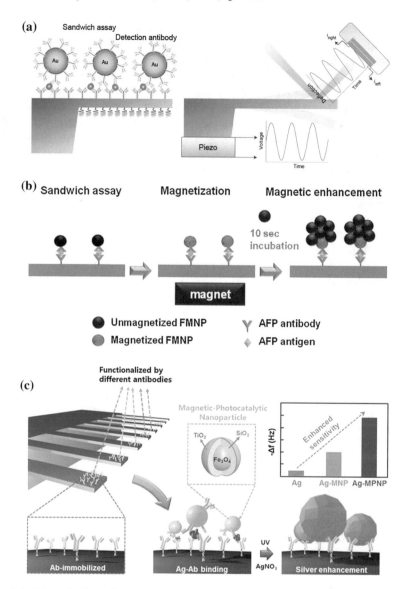

**Fig. 5.6** Schematic illustrations of **a** a sandwich assay on the cantilever (left) and the effect of the sandwich assay on the resonance frequency of the cantilever (right) (Reprinted with permission from Ref. [90]. Copyright 2014 Nature Publishing Group); **b** the FMNP-based sandwich immunoassay and magnetic enhancement reactions (Reprinted with permission from Ref. [93]. Copyright 2012 Royal Society of Chemistry); **c** a hybrid nanoparticle-based sandwich immunoassay with photocatalytic reduction of silver. The free end of each microcantilever was functionalized with different antibodies, and the binding of antibody-conjugated hybrid nanoparticles to the cantilevers as well as the subsequent photocatalytic deposition of silver increases the mass of the nanoparticles, enhancing the sensitivity of the microcantilever resonators (Reprinted with permission from Ref. [94]. Copyright 2012 American Chemical Society)

was subsequently improved to 10 pg mL$^{-1}$ by introducing a sandwich assay that containing detective antibody (anti-GPC1) grafted on 100 nm AuNPs into the cantilever [91]. A similar strategy has been utilized in the nanosensor developed by Kosaka et al. for mechanical and optical detection of CEA and PSA in serum (Fig. 5.6a) [90]. In this report, the biomarker was successively recognized by a surface-immobilized antibody and an unbonded antibody in solution which has been tethered to an AuNP that acts as a plasmonic and mass label and could identify the free region of the captured biomarker. The two signatures were monitored with a silicon cantilever that serves as an optical cavity which boosts the plasmonic signal from the nanoparticles and as a mechanical resonator for "weighing" the mass of the captured nanoparticles. As an ultrasensitive detection limit ($10^{-16}$ g mL$^{-1}$), accompanied an extremely low false positives/negatives rate ($\sim 10^{-4}$) was achieved, this nanosensor represents a significant step toward overcoming the technical difficulties in diagnosing early cancer directly from blood samples [92].

Same as the AuNPs, ferromagnetic nanoparticles (FMNPs) and magnetic photocatalytic hybrid nanoparticles ($Fe_3O_4@SiO_2@TiO_2$) have also been applied to enhance the signal of microcantilever-based protein sandwich assay. Lee et al. described the first use of agglomerated FMNPs in immunoassays for detecting AFP using the microcantilever sensor [93]. As shown in Fig. 5.6b, the FMNPs were magnetized using a permanent magnet subsequently to the bound of the antibody-conjugated FMNPs to the AFP antigen immobilized on the microcantilever. The resulting cantilever was incubated in a solution containing nonmagnetized and unfunctionalized FMNPs to induce the free FMNPs to agglomerate around the magnetized FMNPs. The deflection and resonance frequency of the cantilever were measured after washing and drying the cantilever. Due to the agglomeration of FMNPs, the enhancement reactions could be completed within 10 s, and the deflection and resonance frequency of the cantilever were more than tenfold greater than that before agglomeration of the free nanoparticles.

Different from the physical agglomeration-based signal enhancement method, Joo et al. reported a gravimetric immunoassay for multiple protein biomarker detections, by coupling magnetic preconcentration with photocatalytic silver enhancement reactions [94]. The cantilevers as well as the $Fe_3O_4@SiO_2@TiO_2$ which could preconcentrate the target proteins (by the $Fe_3O_4$ core) and photocatalytically reduce the silver ions to metallic silver (by the $TiO_2$ shell) were functionalized with antibodies. The resonance frequency changes were produced and amplified by the binding of antibody-conjugated hybrid nanoparticles to the cantilevers and the photocatalytic deposition of silver, respectively (Fig. 5.6c). As a result, protein biomarkers including AFP, IL-6, and IFN-γ could be selectively detected within $\sim 1$ h, with LODs ($\sim 0.1$ pg mL$^{-1}$) lower than the clinical threshold of the biomarkers.

## 5.7 Sandwich Assays Based on SAW and RRS

Besides those aforementioned protein sandwich assays based on fluorescence, electrochemistry, and color change or use of SPR, SERS, GMR, QCM, and microcantilever, there are still some noteworthy protein sandwich assays of other styles in spite of their modest reports.

Lee et al. presented a SAW immunosensor for the detection of cardiac markers including creatine kinase (CK)-MB, MYG, and cTnI (Fig. 5.7a). The target detection antibody–AuNP complexes were firstly formed after the targets in human serum were captured on AuNPs which were pretagged with detection antibodies, and addition of these complexes to the sensor surface that immobilized the capture antibody resulted in sandwich immunoassay format and greatly enhanced the signal. A gold staining method was implemented to further amplify the signal. Results showed that sensitivities for the detection of CK-MB, MYG, and cTnI were 1.1, 16.0 ng mL$^{-1}$, and 50 pg mL$^{-1}$, respectively [95]. Inspired by this approach, Zhang et al. developed another SAW immunosensor to detect CEA in exhaled breath condensate. A linear response within the range of 1–16 ng mL$^{-1}$ CEA as well as a LOD of 1 ng mL$^{-1}$ was obtained [96]. By integrating a microfluidic antibody capture chip, on which an antibody-based sandwich assay took place, with the SAW lysis which could partially or completely disrupt cells from a human cancer cell line, Salehi-Reyhani et al. demonstrated the detection of human tumor suppressor protein p53 in the complex mixture of a crude lysate [97].

**Fig. 5.7** Schematic illustrations of **a** a SAW-based sandwich immunoassay in combination with gold staining (Reprinted with permission from Ref. [95]. Copyright 2011 American Chemical Society); **b** a RRS-based sandwich immunoassay for AFP detection (Reprinted with permission from Ref. [98]. Copyright 2013 Springer-Verlag Wien)

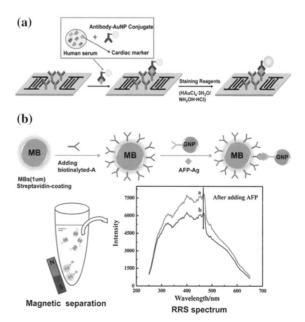

By tagging two antibodies with MBs and AuNPs, respectively, and applying the RRS spectra as the signal output, Lu et al. described a sandwich immunoassay for AFP detection (Fig. 5.7b). The AuNP–antigen–MB complexes were formed in the presence of AFP, and the RRS signal was monitored after separation of the complexes, based on which the AFP could be determined with a LOD of 13.6 pM [98].

## 5.8 Conclusion

This chapter summarized the main recent advances on sandwich assay for protein detection which utilized signal output besides fluorescence, electrochemistry, and color. Through combining the most commonly used recognition molecules, antibodies, and aptamers, with various sensing platforms, different proteins of interest have been sensitively and specifically detected at ultralow concentration levels. Great promises are held by these assays in diverse applications such as medical diagnostics, therapy, clinical research, and basic science; however, room remains to be improved which lies in the following aspects: (1) As the real samples are more complex than those experimental ones, much effort should be put on making these assays more robust and practical; (2) though high sensitivity and specificity have been shown by the antibodies, it should not be ignored that denaturation of proteins is easy but renaturation of them is hard, and thus, more new recognition molecules own stable characteristics such as aptamers, locked nucleic acids, peptide nucleic acids, and even exotic artificial nucleic acids that are urgently to be explored; (3) it is reasonable to believe that integration of existing techniques as well as invention of new strategy will eventually help sandwich assay-based protein detections serving their proper purpose in practical use.

## References

1. Rothenhäusler B, Knoll W (1988) Surface-plasmon microscopy. Nature 332:615–617
2. Yanase Y, Hiragun T, Yanase T, Kawaguchi T, Ishii K, Hide M (2013) Application of SPR imaging sensor for detection of individual living cell reactions and clinical diagnosis of type I allergy. Allergol Int 62:163–169
3. Shabani A, Tabrizian M (2013) Design of a universal biointerface for sensitive, selective, and multiplex detection of biomarkers using surface plasmon resonance imaging. Analyst 138:6052–6062
4. D'Agata R, Spoto G (2013) Surface plasmon resonance imaging for nucleic acid detection. Anal Bioanal Chem 405:573–584
5. Wong CL, Olivo M (2014) Surface plasmon resonance imaging sensors: a review. Plasmonics 9:809–824
6. Zeng S, Baillargeat D, Ho HP, Yong KT (2014) Nanomaterials enhanced surface plasmon resonance for biological and chemical sensing applications. Chem Soc Rev 43:3426–3452

7. Altintas Z, Uludag Y, Gurbuz Y, Tothill IE (2011) Surface plasmon resonance based immunosensor for the detection of the cancer biomarker carcinoembryonic antigen. Talanta 86:377–383
8. Alleyne CJ, Kirk AG, McPhedran RC, Nicorovici NAP, Maystre D (2007) Enhanced SPR sensitivity using periodic metallic structures. Opt Express 15:8163–8169
9. Law WC, Yong KT, Baev A, Prasad PN (2011) Sensitivity improved surface plasmon resonance biosensor for cancer biomarker detection based on plasmonic enhancement. ACS Nano 5:4858–4864
10. Pawula M, Altintas Z, Tothill IE (2016) SPR detection of cardiac troponin T for acute myocardial infarction. Talanta 146:823–830
11. Gnedenko OV, Mezentsev YV, Molnar AA, Lisitsa AV, Ivanov AS, Archakov AI (2013) Highly sensitive detection of human cardiac myoglobin using a reverse sandwich immunoassay with a gold nanoparticle-enhanced surface plasmon resonance biosensor. Anal Chim Acta 759:105–109
12. Lu JD, Van Stappen T, Spasic D, Delport F, Vermeire S, Gils A, Lammertyn J (2016) Fiber optic-SPR platform for fast and sensitive infliximab detection in serum of inflammatory bowel disease patients. Biosens Bioelectron 79:173–179
13. Wu Q, Li S, Sun Y, Wang JN (2017) Hollow gold nanoparticle-enhanced SPR based sandwich immunoassay for human cardiac troponin I. Microchim Acta 184:2395–2402
14. Wang Y, Dostalek J, Knoll W (2011) Magnetic nanoparticle-enhanced biosensor based on grating-coupled surface plasmon resonance. Anal Chem 83:6202–6207
15. Guo XW (2014) $Fe_3O_4$@Au nanoparticles enhanced surface plasmon resonance for ultrasensitive immunoassay. Sens Actuators B: Chem 205:276–280
16. Jatschka J, Dathe A, Csáki A, Fritzsche W, Stranik O (2016) Propagating and localized surface plasmon resonance sensing-A critical comparison based on measurements and theory. Sens Bio-Sens Res 7:62–70
17. Wu B, Jiang R, Wang Q, Huang J, Yang XH, Wang KM, Li WS, Chen ND, Li Q (2016) Detection of C-reactive protein using nanoparticle-enhanced surface plasmon resonance using an aptamer-antibody sandwich assay. Chem Commun 52:3568–3571
18. Bai YF, Feng F, Zhao L, Wang CY, Wang HY, Tian MZ, Qin J, Duan Y, He XX (2013) Aptamer/thrombin/aptamer-AuNPs sandwich enhanced surface plasmon resonance sensor for the detection of subnanomolar thrombin. Biosens Bioelectron 47:265–270
19. Tan WH, Donovan MJ, Jiang JH (2013) Aptamers from cell-based selection for bioanalytical applications. Chem Rev 113:2842–2862
20. Sim HR, Wark AW, Lee HJ (2010) Attomolar detection of protein biomarkers using biofunctionalized gold nanorods with surface plasmon resonance. Analyst 135:2528–2532
21. Kim S, Lee HJ (2015) Direct detection of α-1 antitrypsin in serum samples using surface plasmon resonance with a new aptamer-antibody sandwich assay. Anal Chem 87:7235–7240
22. Mir M, Vreeke M, Katakis I (2006) Different strategies to develop an electrochemical thrombin aptasensor. Electrochem Commun 8:505–511
23. Vance SA, Sandros MG (2014) Zeptomole detection of C-reactive protein in serum by a nanoparticle amplified surface plasmon resonance imaging aptasensor. Sci Rep 4:5129
24. Nguyen VT, Seo HB, Kim BC, Kim SK, Song CS, Gu MB (2016) Highly sensitive sandwich-type SPR based detection of whole H5Nx viruses using a pair of aptamers. Biosens Bioelectron 86:293–300
25. Haes AJ, Chang L, Klein WL, Van Duyne RP (2005) Detection of a biomarker for Alzheimer's disease from synthetic and clinical samples using a nanoscale optical biosensor. J Am Chem Soc 127:2264–2271
26. Guo LH, Kim DH (2012) LSPR biomolecular assay with high sensitivity induced by aptamer-antigen-antibody sandwich complex. Biosens Bioelectron 31:567–570
27. Stewart ME, Anderton CR, Thompson LB, Maria J, Gray SK, Rogers JA, Nuzzo RG (2008) Nanostructured plasmonic sensors. Chem Rev 108:494–521
28. Homola J (2008) Surface plasmon resonance sensors for detection of chemical and biological species. Chem Rev 108:462–493

29. Kabashin AV, Evans P, Pastkovsky S, Hendren W, Wurtz GA, Atkinson R, Pollard R, Podolskiy VA, Zayats AV (2009) Plasmonic nanorod metamaterials for biosensing. Nat Mater 8:867–871

30. Yuan YF, Panwar N, Yap SHK, Wu Q, Zeng SW, Xu JH, Tjin SC, Song J, Qu J, Yong KT (2017) SERS-based ultrasensitive sensing platform: An insight into design and practical applications. Coordin Chem Rev 337:1–33

31. Wang ZY, Zong SF, Wu L, Zhu D, Cui YP (2017) SERS-activated platforms for immunoassay: probes, encoding methods, and applications. Chem Rev 117:7910–7963

32. Han XX, Zhao B, Ozaki Y (2009) Surface-enhanced Raman scattering for protein detection. Anal Bioanal Chem 394:1719–1727

33. Burda C, Chen X, Narayanan R, El-Sayed MA (2005) Chemistry and properties of nanocrystals of different shapes. Chem Rev 105:1025–1102

34. Annadhasan M, Muthukumarasamyvel T, Sankar Babu VR, Rajendiran N (2014) Green synthesized silver and gold nanoparticles for colorimetric detection of $Hg^{2+}$, $Pb^{2+}$, and $Mn^{2+}$ in aqueous medium. ACS Sustain Chem Eng 2:887–896

35. Wang Y, Tang LJ, Jiang JH (2013) Surface-enhanced Raman spectroscopy-based, homogeneous, multiplexed immunoassay with antibody-fragments-decorated gold nanoparticles. Anal Chem 85:9213–9220

36. Fu XL, Chen LX, Choo J (2017) Optical nanoprobes for ultrasensitive immunoassay. Anal Chem 89:124–137

37. Han XX, Kitahama Y, Itoh T, Wang CX, Zhao B, Ozaki Y (2009) Protein-mediated sandwich strategy for surface-enhanced Raman scattering: application to versatile protein detection. Anal Chem 81:3350–3355

38. Chon H, Lim C, Ha SM, Ahn Y, Lee EK, Chang SI, Seong GH, Choo J (2010) On-chip immunoassay using surface-enhanced Raman scattering of hollow gold nanospheres. Anal Chem 82:5290–5295

39. Li M, Cushing SK, Zhang JM, Suri S, Evans R, Petros WP, Gibson LF, Ma DL, Liu YX, Wu NQ (2013) Three-dimensional hierarchical plasmonic nano-architecture enhanced surface-enhanced Raman scattering immunosensor for cancer biomarker detection in blood plasma. ACS Nano 7:4967–4976

40. Zeman EJ, Schatz GC (1987) An accurate electromagnetic theory study of surface enhancement factors for silver, gold, copper, lithium, sodium, aluminum, gallium, indium, zinc, and cadmium. J Phys Chem 91:634–643

41. Hwang H, Chon H, Choo J, Park JK (2010) Optoelectrofluidic sandwich immunoassays for detection of human tumor marker using surface-enhanced Raman scattering. Anal Chem 82:7603–7610

42. Liang JJ, Liu HW, Huang CH, Yao CZ, Fu QQ, Li XQ, Cao DL, Luo Z, Tang Y (2015) Aggregated silver nanoparticles based surface-enhanced Raman scattering enzyme-linked immunosorbent assay for ultrasensitive detection of protein biomarkers and small molecules. Anal Chem 87:5790–5796

43. Wang YQ, Yan B, Chen LX (2013) SERS tags: novel optical nanoprobes for bioanalysis. Chem Rev 113:1391–1428

44. Guarrotxena N, Bazan GC (2014) Antitags: SERS-encoded nanoparticle assemblies that enable single-spot multiplex protein detection. Adv Mater 26:1941–1946

45. Fabris L, Dante M, Nguyen T-Q, Tok JBH, Bazan GC (2008) SERS aptatags: new responsive metallic nanostructures for heterogeneous protein detection by surface enhanced Raman spectroscopy. Adv Funct Mater 18:2518–2525

46. Zengin A, Tamer U, Caykara T (2015) Fabrication of a SERS based aptasensor for detection of ricin B toxin. J Mater Chem B 3:306–315

47. Ye J, Chen Y, Liu Z (2014) A boronate affinity sandwich assay: an appealing alternative to immunoassays for the determination of glycoproteins. Angew Chem Int Ed 53:10386–10389

48. Yang X, Gu C, Qian F, Li Y, Zhang JZ (2011) Highly sensitive detection of proteins and bacteria in aqueous solution using surface-enhanced Raman scattering and optical fibers. Anal Chem 83:5888–5894

49. Zeng H, Sun SH (2008) Syntheses, properties, and potential applications of multicomponent magnetic nanoparticles. Adv Funct Mater 18:391–400
50. Chon H, Lee S, Son SW, Oh CH, Choo J (2009) Highly sensitive immunoassay of lung cancer marker carcinoembryonic antigen using surface-enhanced Raman scattering of hollow gold nanospheres. Anal Chem 81:3029–3034
51. Cheng ZY, Choi N, Wang R, Lee S, Moon KC, Yoon SY, Chen LX, Choo J (2017) Simultaneous detection of dual prostate specific antigens using surface-enhanced Raman scattering-based immunoassay for accurate diagnosis of prostate cancer. ACS Nano 11:4926–4933
52. Li J, Skeete Z, Shan SY, Yan S, Kurzatkowska K, Zhao W, Ngo QM, Holubovska P, Luo J, Hepel M, Zhong CJ (2015) Surface enhanced Raman scattering detection of cancer biomarkers with bifunctional nanocomposite probes. Anal Chem 87:10698–10702
53. Song CY, Yang YJ, Yang BY, Min LH, Wang LH (2016) Combination assay of lung cancer associated serum markers using surface-enhanced Raman spectroscopy. J Mater Chem B 4:1811–1817
54. Liu H-L, Nosheen F, Wang X (2015) Noble metal alloy complex nanostructures: controllable synthesis and their electrochemical property. Chem Soc Rev 44:3056–3078
55. Kong XM, Yu Q, Zhang XF, Du XZ, Gong H, Jiang H (2012) Synthesis and application of surface enhanced Raman scattering (SERS) tags of Ag@SiO$_2$ core/shell nanoparticles in protein detection. J Mater Chem 22:7767–7774
56. Hu PP, Liu H, Zhan L, Zheng LL, Huang CZ (2015) Coomassie brilliant blue R-250 as a new surface-enhanced Raman scattering probe for prion protein through a dual-aptamer mechanism. Talanta 139:35–39
57. Wu L, Wang ZY, Fan KQ, Zong SF, Cui YP (2015) A SERS-assisted 3d barcode chip for high-throughput biosensing. Small 11:2798–2806
58. Baibich MN, Broto JM, Fert A, Van Dau FN, Petroff F, Etienne P, Creuzet G, Friederich A, Chazelas J (1988) Giant magnetoresistance of (001)Fe/(001)Cr magnetic superlattices. Phys Rev Lett 61:2472–2475
59. Binasch G, Grünberg P, Saurenbach F, Zinn W (1989) Enhanced magnetoresistance in layered magnetic structures with antiferromagnetic interlayer exchange. Phys Rev B 39:4828–4830
60. Cubells-Beltrán MD, Reig C, Madrenas J, De Marcellis A, Santos J, Cardoso S, Freitas P (2016) Integration of GMR sensors with different technologies. Sens-Basel 16:939
61. Reig C, Cardoso S, Mukhopadhyay S (2013) Giant magnetoresistance (GMR) sensors: from basis to state-of-the-art applications. Springer, Heidelberg
62. Fishbein I, Levy RJ (2009) Analytical chemistry: the matrix neutralized. Nature 461:890–891
63. Krishna VD, Wu K, Perez AM, Wang JP (2016) Giant magnetoresistance-based biosensor for detection of influenza A virus. Front Microbiol 7:400
64. Shen JW, Li YB, Gu HS, Xia F, Zuo XL (2014) Recent development of sandwich assay based on the nanobiotechnologies for proteins, nucleic acids, small molecules, and ions. Chem Rev 114:7631–7677
65. Wang Y, Wang W, Yu L, Tu L, Feng YL, Klein T, Wang JP (2015) Giant magnetoresistive-based biosensing probe station system for multiplex protein assays. Biosens Bioelectron 70:61–68
66. Wang T, Yang Z, Lei C, Lei J, Zhou Y (2014) An integrated giant magnetoimpedance biosensor for detection of biomarker. Biosens Bioelectron 58:338–344
67. Gaster RS, Hall DA, Nielsen CH, Osterfeld SJ, Yu H, Mach KE, Wilson RJ, Murmann B, Liao JC, Gambhir SS, Wang SX (2009) Matrix-insensitive protein assays push the limits of biosensors in medicine. Nat Med 15:1327–1332
68. Srinivasan B, Li YP, Jing Y, Xu YH, Yao XF, Xing CG, Wang JP (2009) A detection system based on giant magnetoresistive sensors and high-moment magnetic nanoparticles demonstrates zeptomole sensitivity: potential for personalized medicine. Angew Chem Int Ed 48:2764–2767

69. Li YP, Srinivasan B, Jing Y, Yao XF, Hugger MA, Wang JP, Xing CG (2010) Nanomagnetic competition assay for low-abundance protein biomarker quantification in unprocessed human sera. J Am Chem Soc 132:4388–4392

70. Srinivasan B, Li YP, Jing Y, Xing CG, Slaton J, Wang JP (2011) A three-layer competition-based giant magnetoresistive assay for direct quantification of endoglin from human urine. Anal Chem 83:2996–3002

71. Ivanov K, Kolev N, Tonev A, Nikolova G, Krasnaliev I, Softova E, Tonchev A (2009) Comparative analysis of prognostic significance of molecular markers of apoptosis with clinical stage and tumor differentiation in patients with colorectal cancer: a single institute experience. Hepato-Gastroentero 56:94–98

72. Madu CO, Lu Y (2010) Novel diagnostic biomarkers for prostate cancer. J Cancer 1:150–177

73. Yang Z, Liu Y, Lei C, Sun XC, Zhou Y (2015) A flexible giant magnetoimpedance-based biosensor for the determination of the biomarker C-reactive protein. Microchim Acta 182:2411–2417

74. Osterfeld SJ, Yu H, Gaster RS, Caramuta S, Xu L, Han SJ, Hall DA, Wilson RJ, Sun SH, White RL, Davis RW, Pourmand N, Wang SX (2008) Multiplex protein assays based on real-time magnetic nanotag sensing. Proc Natl Acad Sci USA 105:20637–20640

75. Mak AC, Osterfeld SJ, Yu H, Wang SX, Davis RW, Jejelowo OA, Pourmand N (2010) Sensitive giant magnetoresistive-based immunoassay for multiplex mycotoxin detection. Biosens Bioelectron 25:1635–1639

76. Ng E, Nadeau KC, Wang SX (2016) Giant magnetoresistive sensor array for sensitive and specific multiplexed food allergen detection. Biosens Bioelectron 80:359–365

77. Choi J, Gani AW, Bechstein DJB, Lee JR, Utz PJ, Wang SX (2016) Portable, one-step, and rapid GMR biosensor platform with smartphone interface. Biosens Bioelectron 85:1–7

78. He P, Liu LJ, Qiao WP, Zhang SS (2014) Ultrasensitive detection of thrombin using surface plasmon resonance and quartz crystal microbalance sensors by aptamer-based rolling circle amplification and nanoparticle signal enhancement. Chem Commun 50:1481–1484

79. Cheng CI, Chang YP, Chu YH (2012) Biomolecular interactions and tools for their recognition: focus on the quartz crystal microbalance and its diverse surface chemistries and applications. Chem Soc Rev 41:1947–1971

80. Cooper MA, Singleton VT (2007) A survey of the 2001 to 2005 quartz crystal microbalance biosensor literature: applications of acoustic physics to the analysis of biomolecular interactions. J Mol Recognit 20:154–184

81. Henne WA, Doorneweerd DD, Lee J, Low PS, Savran C (2006) Detection of folate binding protein with enhanced sensitivity using a functionalized quartz crystal microbalance sensor. Anal Chem 78:4880–4884

82. Lei JP, Ju HX (2012) Signal amplification using functional nanomaterials for biosensing. Chem Soc Rev 41:2122–2134

83. Chen Q, Tang W, Wang DZ, Wu XJ, Li N, Liu F (2010) Amplified QCM-D biosensor for protein based on aptamer-functionalized gold nanoparticles. Biosens Bioelectron 26:575–579

84. Ogi H, Yanagida T, Hirao M, Nishiyama M (2011) Replacement-free mass-amplified sandwich assay with 180-MHz electrodeless quartz-crystal microbalance biosensor. Biosens Bioelectron 26:4819–4822

85. Mashaghi A, Mashaghi S, Reviakine I, Heeren RMA, Sandoghdar V, Bonn M (2014) Label-free characterization of biomembranes: from structure to dynamics. Chem Soc Rev 43:887–900

86. Alfonta L, Willner I, Throckmorton DJ, Singh AK (2001) Electrochemical and quartz crystal microbalance detection of the cholera toxin employing horseradish peroxidase and GM1-functionalized liposomes. Anal Chem 73:5287–5295

87. Deng Y, Yue XL, Hu H, Zhou XD (2017) A new analytical experimental setup combining quartz crystal microbalance with surface enhancement Raman spectroscopy and its application in determination of thrombin. Microchem J 132:385–390

88. Uludag Y, Tothill IE (2012) Cancer biomarker detection in serum samples using surface plasmon resonance and quartz crystal microbalance sensors with nanoparticle signal amplification. Anal Chem 84:5898–5904
89. Zhu Q (2011) Microcantilever sensors in biological and chemical detections. Sens transducers 125:1–21
90. Kosaka PM, Pini V, Ruz JJ, da Silva RA, González MU, Ramos D, Calleja M, Tamayo J (2014) Detection of cancer biomarkers in serum using a hybrid mechanical and optoplasmonic nanosensor. Nat Nano 9:1047–1053
91. Etayash H, McGee AR, Kaur K, Thundat T (2016) Nanomechanical sandwich assay for multiple cancer biomarkers in breast cancer cell-derived exosomes. Nanoscale 8:15137–15141
92. Longo G (2014) Cancer biomarkers: detected twice for good measure. Nat Nano 9:959–960
93. Lee D, Kwon D, Ko W, Joo J, Seo H, Lee SS, Jeon S (2012) A rapid and facile signal enhancement method for microcantilever-based immunoassays using the agglomeration of ferromagnetic nanoparticles. Chem Commun 48:7182–7184
94. Joo J, Kwon D, Yim C, Jeon S (2012) Highly sensitive diagnostic assay for the detection of protein biomarkers using microresonators and multifunctional nanoparticles. ACS Nano 6:4375–4381
95. Lee J, Choi YS, Lee Y, Lee HJ, Lee JN, Kim SK, Han KY, Cho EC, Park JC, Lee SS (2011) Sensitive and simultaneous detection of cardiac markers in human serum using surface acoustic wave immunosensor. Anal Chem 83:8629–8635
96. Zhang X, Zou YC, An C, Ying KJ, Chen X, Wang P (2015) Sensitive detection of carcinoembryonic antigen in exhaled breath condensate using surface acoustic wave immunosensor. Sens Actuators B: Chem 217:100–106
97. Salehi-Reyhani A, Gesellchen F, Mampallil D, Wilson R, Reboud J, Ces O, Willison KR, Cooper JM, Klug DR (2015) Chemical-free lysis and fractionation of cells by use of surface acoustic waves for sensitive protein assays. Anal Chem 87:2161–2169
98. Lu Y, Huang XY, Ren JC (2013) Sandwich immunoassay for alpha-fetoprotein in human sera using gold nanoparticle and magnetic bead labels along with resonance Rayleigh scattering readout. Microchim Acta 180:635–642

# Chapter 6
# Colorimetric Sandwich Assays for Nucleic Acid Detection

Xiaoxia Hu and Quan Yuan

**Abstract** Colorimetric sandwich assay has attracted wide interests in detection of nucleic acid due to the advantages of convenience and visibility. In this chapter, we summarized the development of colorimetric sandwich assay employed nanoparticles-, traditional enzyme-, and DNAzyme-based sensor, aiming at providing a general guide for designing colorimetric sandwich assay for the detection of nucleic acid. Furthermore, we discussed the challenges in the development of colorimetric sandwich assay regarding sensitivity and stability, thus offering further opportunities to develop more of robust colorimetric sandwich assay for the detection of nucleic acid.

**Keywords** Colorimetric sandwich assay · Nucleic acid · Detection
Nanoparticles · Traditional enzyme · DNAzyme · Sensitivity · Stability

## 6.1 Introduction

The detection of nucleic acids is fundamental for investigating their functions and developing molecular diagnostics for clinical application. For decades, the sandwich assay has been a promising strategy for the detection of nucleic acids. The basic strategy of the nucleic acid sandwich assay is the nucleic acid hybridization technique, which can be used for the analysis of specific nucleic acid sequences. The sandwich assays usually need the simultaneous binding of two kinds of specific nucleic acid-based probe, thus making them extremely specific. In addition, the amplified signaling mechanism is often coupled into the basic sandwich architecture, thus achieving impressive sensitivity detection.

X. Hu · Q. Yuan (✉)
College of Chemistry and Molecular Sciences, Wuhan University, Wuhan 430072, People's Republic of China
e-mail: yuanquan@whu.edu.cn

X. Hu
e-mail: xioaxiajiay@whu.edu.cn

© Springer Nature Singapore Pte Ltd. 2018
F. Xia et al. (eds.), *Biosensors Based on Sandwich Assays*,
https://doi.org/10.1007/978-981-10-7835-4_6

The ability to detect the specific DNA sequences in a convenient and inexpensive way is important in clinical diagnostics. In colorimetric assay, the color change can be observed with the naked eye upon the addition of analytes, which holds great promise for inexpensive and real-time detection of target DNA sequence. Therefore, colorimetric sandwich assay is likely to provide a promising tool for fundamental research and clinic diagnosis with expected convenience, visibility, promoting the utilization of real-time assay of analyte. Particularly, the colorimetric sandwich assay employed with nanoparticles- [1–3], traditional enzyme- [4–6], and DNAzyme-based sensor [7–10] has recently attracted considerable attention in the diagnostic applications thanks to their versatility and simplicity. Herein, we present a critical review of the literature on the colorimetric nucleic acid sandwich assay to summarize and comment upon its development, advances, and challenges.

## 6.2  Colorimetric Sandwich Assays Based on Gold Nanoparticles

Nanomaterials possess unique optical properties in comparison to bulk materials. One of the most popular nanomaterials applied for colorimetric assay is gold nanoparticles (AuNPs), which possess distinct distance-dependent optical properties due to the interparticle plasmon coupling [2]. When the AuNPs get into proximity with one another, its scattering profile can be changed and thus lead to a pronounced color change with a shift of absorption spectrum [11–13]. Taking the advantage of the color change derived from the interparticle plasmon coupling during AuNPs aggregation (red-to-blue or purple) or redispersion of an AuNPs aggregates (blue or purple-to-red), the AuNPs-based colorimetric assay shows a great interest in visual detection of a wide variety of targets. Classically, sandwich-based strategy has been applied in colorimetric assay to detect the nucleic acid in vitro and in vivo [14, 15]. Such a colorimetric sandwich assay can achieve impressive specificity and sensitivity of detection due to the specific binding between antibody–antigen pairs and the signal amplify. The foremost research on colorimetric sandwich assay for nucleic acid was reported by the Mirkin group [16, 17]. The colorimetric-based sensing platform for DNA detection with high selectivity can be achieved due to the formation of a network of AuNPs-based nanoprobes with a concomitant color change. When the interparticle distance of the AuNPs aggregates is substantially greater than the average particle diameter, the sensing platform appears red color. Conversely, when the interparticle distance decrease to less than the average particle diameter, the color of the sensing platform changes to blue. Such a color change is attributed to the surface plasmon resonance of AuNPs. Based on this unique phenomenon, Mirkin group developed a colorimetric sandwich assay which is sensitive, selective, visual, and instrument-free for DNA detection [16]. Two kinds of thiol tether linked probes, which are used for one oligonucleotide target, are covalently connected to the surface of AuNPs with an average diameter of 13 nm. Each capture DNA linked on the probe possesses 28 bases, the first 13 nucleotides serve as a flexible spacer, and the last 15 act as a recognition fragment for the target. The recognition segments of two kinds of probes can align contiguously on the target through the hybridization. As a result, color of the solution changed from red to purple with the introduction of target. However, it needed a standing overnight to finish the hybridization. The slow reaction rate of hybridization was due to the steric considerations and high negative charge density on the surface of AuNPs. This slow reaction kinetics can be improved significantly when the solution was heated to 50 °C for 5 min or freezed in a bath of dry ice and isopropyl alcohol and then thawed at room temperature. They found that the transition required the existence of all three components which included two kinds of probes and the target. Through the denaturation of the hybridized aggregates, the melting curve for the nanoparticle system is sharper than that for the system without nanoparticles. On the basis of this mechanism, this three-component nanoparticle-based colorimetric sandwich assay is more selective than any two-component assay system employed a single-strand oligonucleotide hybridizing with the target. The colorimetric sandwich assay has an optimistic application in detection with simplicity in instrumentation and operation.

Since lots of secondary structures in nucleic acid targets severely hinder the hybridization with the complementary oligonucleotide probes, the sensitivity of the hybridization-based detection can be decreased [18]. To make the hybridization-based assay of nucleic acids more efficient, it needs a stringent control over the detection conditions. The low strict conditions such as high salt concentration and/or low temperature is beneficial to form the target-probe duplexes, but the secondary structure in the nucleic acid target remains stabilized, making the complementary base-pairing reaction between the probe and target largely inaccessible. The high strict conditions such as low salt concentration and/or high temperature decrease secondary structure. However, such a stringent condition could make the target-probe duplexes less stable, leading to weak signals [19, 20]. The unavoidable presence of secondary structure seriously reduces the reliability of the hybridization-based assays. To explore the thermal stability of the DNA duplexes, Liu's group [21] studied DNA-linked AuNPs in many kinds of ionic liquids (ILs). They found that ILs can not only serve as cations to screen DNA charges, but also can solvate DNA bases. For alkylammonium nitrate ILs, it is not accurate to perform the traditional DNA hyperchromicity experiment. However, the AuNPs-based sandwich assay can still be successfully conducted. So, the DNA-functionalized AuNPs can still maintain their functions in ILs. ILs were demonstrated to be useful for AuNPs-based colorimetric sandwich DNA assay. To alleviate the secondary structure of target DNA, Gao et al. [22] reported a method to detect the DNA under extremely low salt conditions (Fig. 6.1). In such a condition, the secondary structures of DNA are less stable and more accessible. They functionalized the nonionic morpholino oligos on the AuNPs to construct a new type of nanoparticle probes. Due to the salt-independent hybridization of probes with DNA targets, the sandwich DNA assay for colorimetric target recognition can be successfully performed in a low salt condition. The study exhibits superior performance in DNA assay bearing secondary structure and improves the hybridization efficiency.

To improve the detection sensitivity, Mirkin group further developed a three-component sandwich assay for DNA detection [17]. The 3'-thiol-modified capture DNA was firstly immobilized on the float glass microscope slide to construct a target-active substrate. Nanoparticle probes and targets were then cohybridized to the target-active substrate. When the target concentration was above 1 nM, the substrate surface appeared light pink due to the high density of the hybridized AuNPs. However, the attached AuNPs could not be visualized with the lower target concentrations ($\leq 100$ pM). To facilitate the visualization of the nanoparticles hybridized substrate, they applied the reduction of silver ions to realize the signal amplification. Silver metal was deposited on the surface of the AuNPs by employing hydroquinone as the reducing agent for silver ion. Improvement of two orders of magnitude was achieved over the fluorophore-based sandwich assay. Such a silver amplification system inherently changed the melting profile of the targets, which permits the excellent differentiation of single-nucleotide mismatch.

Another strategy, which uses the enzymatic ligation chain reaction (LCR) concept, is applied to further improve the detection sensitivity [23]. The target DNA first hybridized with two types of the capture probe DNA-coated AuNPs (CP-coated AuNPs) to form DNA duplexes. Subsequently, ampligase ligated two kinds of CP-coated AuNPs templated by the target DNA. When increasing the temperature to 90 °C, the duplexes were denatured and thus releasing the target DNA and the ligated AuNPs. By repeating the hybridization and ligation, the number of ligated AuNPs was amplified, and eventually resulted in the color change of the solution from red to purple (Fig. 6.2). Since the ligated AuNPs can be amplified with the thermal cycling, ultrasensitive (aM level) can be realized.

Besides the detection sensitivity, stability is another important issue for AuNPs-based colorimetric assays [24, 25]. The capture probe DNA modified on AuNPs is often unstable in real samples, in which high concentration of salts could

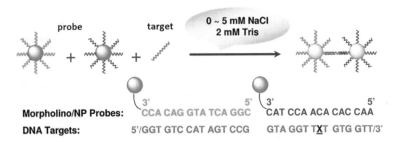

**Fig. 6.1** Schematic presentation of the AuNPs-based sandwich assay of DNA under low salt conditions (Reprinted with permission from Ref. [22]. Copyright 2011 American Chemical Society)

**Fig. 6.2** Target DNA was hybridized with two types of capture DNA-coated AuNPs to form DNA duplexes, and then the two AuNPs was linked together by the ampligase. The target DNA, the ligated AuNPs, and the partial probes were released after denaturation at 90 °C. Then the process of hybridization, ligation, and denaturation was repeated, resulting in the color change (Reprinted with permission from Ref. [23]. Copyright 2012 American Chemical Society)

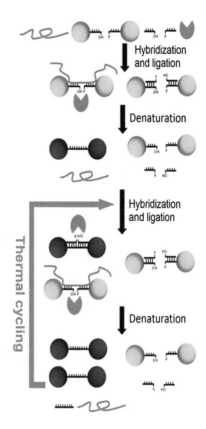

induce the AuNPs aggregation, thus inducing nonspecific signals. Additionally, the biothiols in real samples tend to replace the capture probe DNA modified on AuNPs and thus decrease the detection ability of the AuNPs-based assays. The AuNPs aggregates are unstable in solution, and the color of the aggregation solution finally became colorless in few hours to days. To solve this problem, Kim's group [26] developed an AuNPs-based colorimetric assay with high stability and sensitivity. The oriented hybridization of asymmetrically functionalized AuNPs leaded to form dimers, where the spacing between interparticles was shorter than 1 nm. Since the short pacing in dimers, the obvious color change from red to blue was obtained (Fig. 6.3). The asymmetrically PEGylated AuNPs make the assay very stability in complex samples, thus avoiding the generation of false positive signals. This modification also prevents the generation of multimers and aggregates.

**Fig. 6.3 a** AuNPs were immobilized onto a CTAB bilayer-modified glass substrate. After exposure to the solution containing PEG-thiols, followed by sonication in distilled water, the partial PEGylated AuNPs were obtained. **b** In the presence of the target, perfect complementary Y-shaped DNA duplex was formed (Reprinted with permission from Ref. [26]. Copyright 2013 American Chemical Society)

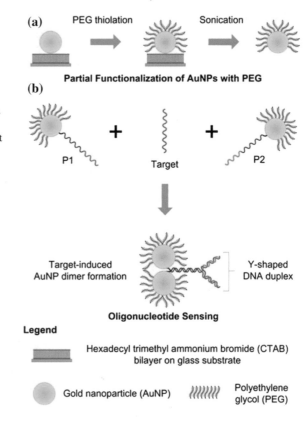

## 6.3 Colorimetric Sandwich Assays Based on Traditional Enzymes

Enzymes have been widely applied for colorimetric detection of targets since they can provide recognition and amplification of the binding event with a visual signal output [27, 28]. Specifically, traditional enzymes are successfully used in colorimetric DNA sandwich assay [5, 29]. The commonly used enzymes include horseradish peroxidase (HRP), lipase, glucose oxidase, and alkaline phosphatase. HRP can be efficiently combined onto the DNA strand through electrostatic interactions, and it can give an easy-to-read color change after target binding. Therefore, HRP has been used as the popular traditional enzyme in colorimetric DNA analysis. Taking the advantage of electrostatic interactions between negative backbone of DNA and HRP, Gao and his colleagues have developed a HRP-based label-free assay for nucleic acids [30]. The neutral peptide nucleic acid (PNA) was chose as a capture probe, and it was immobilized on a streptavidin (SA)-coated plate. The target DNA could hybridize with the capture DNA, which brought in a high density of negative charges. HRP can be adsorbed onto the DNA hybridization under a proper pH condition. The colorimetric detection can be finally realized since the adsorbed HRP catalyzed the color reaction of 3,3',5,5'-Tetramethylbenzidine (TMB) and $H_2O_2$. This proposed assay has further been applied for detecting total RNA samples extracted from human cancer cell lines. Through the extensive adsorption of HRP, the detection limit was down to 0.1–0.2 nM for DNA.

The application of PNA probes for DNA detection is limited due to the short of backbone modifications that can be employed to adjust the basic properties of probe sequences. To address this issue, Appella's group developed a cyclopentane groups-modified PNA for colorimetric sandwich-hybridization DNA assay [31]. In this strategy, one PNA was used as a capture probe and was covalently attached to a plate. Another PNA was modified with cyclopentane groups and was applied as a detection probe. In the presence of target DNA, the sandwich complex formed and the HRP-avidin conjugate could be bound with cyclopentane-labeled PNA. The substrate of tetramethylbenzidine (TMB) could be oxidized and thus generating a colorimetric signal. This sandwich assay can improve the melting temperature and specificity for complementary DNA. Importantly, cyclopentane groups used in this system can improve the detection sensitivity and selectivity. By doing so, the detection limit of DNA was down to 10 zmol, and two cell lines of anthrax have been obviously distinguished through the visible color change.

To integrate multiple functions (e.g., recognition, amplification) into a colorimetric sandwich assay, Fan and his coworkers have designed a "sandwich-type" detection strategy contained multi-component DNA detection nanoprobes for DNA assay [32]. The biotinylated capture probe and thiolated detection probe were separately loaded on the magnetic microparticles (MMPs) and AuNPs. Specifically, the capture probe and detection probe could be flanked the target DNA sequence through the hybridization reaction. Thus, the target DNA could bring the detection probe-linked AuNPs to the proximity of capture DNA-linked MMPs, and the formed complex could be separated by the magnetic field. The HRP confined on the surface of AuNPs catalytically oxidized the substrate and thus generating optical signals that responded the quantity of target DNA. In this strategy, HRP was applied for signal amplification, which could catalyze >10,000 substrate turnovers. They could efficiently distinguish 100 pM target DNA from the color contrast with the naked eye. By employing the instrument-based assays, the lower concentration of target could be also detected (25 pM with absorption and 1 pM with fluorescence). In addition, the nonspecific signals could be suppressed by using bovine serum albumin (BSA) as a nonspecific blocker, thus decreasing the background. Significantly, this multi-component-based sandwich assay performed well in the complicated biological system such as serum.

To improve the detection sensitivity, Fan's group further developed a colorimetric sandwich assay for optical detection of DNA [33]. A crosslinking probe was integrated into the aforementioned system to develop a multi-functional crosslinked Au aggregates. In the presence of target DNA, the multi-component-based nanoprobe was attached to the MMP, and then the complex was further hybridized with the crosslinking nanoprobe to form a crosslinked aggregate. HRP that immobilized on the surface of AuNPs aggregates could catalyze the substrate and produce an optical signal (Fig. 6.4). Compared with the multi-component-based sandwich assay that reported before, this system was more sensitive since much more HRP was attached on the AuNPs aggregates. As expected, target DNA as low as 1 fM could be visually detected.

The ultrasensitive and selective assay of nucleic acid at low physiological levels is important in life science. In this regard, coupling amplifying techniques [e.g., polymerase chain reaction (PCR), hybridization chain reaction (HCR)] to the colorimetric sandwich assay is promising for improving the detection sensitivity and reliability. Recent advances in HCR have gained comprehensive attention in terms of an isothermal replication process triggered by target DNA. Yang et al. developed a colorimetric detection platform based on the HCR-triggered enzyme cascade amplification for DNA detection [5]. They integrated two different hairpin DNA strands [one modified with glucose oxidase (GOx-$H_1$), another modified with HRP (HRP-$H_2$)] with the target DNA. In the presence of target DNA, the HCR of GOx-$H_1$ and HRP-$H_2$ was triggered and generated the double-stranded DNA frameworks, leading to the multiple cascades of HRP and GOx on the HCR chains. The complementary sequence of target DNA and interlaced complementary stem-loop sequences were formed a sandwich structure with long double DNA strands. The added substrate of glucose was oxidized by GOx and yielded $H_2O_2$ as product. Then the formed $H_2O_2$ acted as the substrate for HRP that oxidized 2, 2'-azino-bis (3-ethylbenzothia-zoline-6-sulfonic acid) ($ABTS^{2-}$) to the colored product of $ABTS^{\bullet-}$. In this regard, the produce rate of $ABTS^{\bullet-}$ was related to the

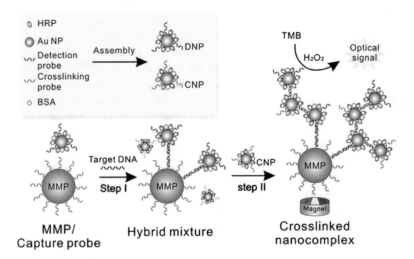

**Fig. 6.4** In the presence of target DNA, the multi-component detection nanoprobe (DNP) was linked to the magnetic microparticles (MMPs) (step I). These complexes were further hybridized with CNP to form a crosslinked aggregate (step II). After magnetic separation, HRP that attached on the surface of Au aggregates could catalyze the enzyme substrate and generate an optical signal (Reprinted with permission from Ref. [33]. Copyright 2009 Elsevier)

concentration of target DNA. This HCR-based colorimetric assay leaded to excellent sensitivity. In addition, the biological substances could be also detected.

## 6.4 Colorimetric Sandwich Assays Based on DNAzyme

In addition to traditional enzyme, nucleic acids that show catalytic properties (DNAzymes) have attracted wide attentions recently [34, 35]. The G-quartet-based DNAzymes is one of the most attractive DNAzymes that formed by a guanine-rich nucleic acid and hemin [36, 37]. This DNAzyme is more stable and robust than traditional enzymes, and it reveals peroxidase-like properties. The peroxidase-mimicking DNAzyme can catalyze the $H_2O_2$-mediated oxidation of ABTS and thus produce the colored radical anion (ABTS$^{\cdot-}$). The DNAzyme has been used as a catalytic label for colorimetric assay in the biosensor and biodetection fields [38–44].

Wang and his coworkers engineered a hemin-G-quartet complex consisting of two guanine-rich single-strand DNAs [41]. The G-quartet structure with two free nucleic acid strands was formed when incubated with hemin. Since this structure has a ternary structure and two free strands, it was unstable and has low catalytic activity. However, when treated with a low level of complementary DNA as target DNA, this structure became more stable, leading to an increased catalytic activity (Fig. 6.5). This hemin-G-quartet supramolecular complex can catalyze the oxidation of ABTS to produce a color change. Thus, the enhanced colorimetric signal originating from the enhanced catalytic DNAzymes could reveal the concentration of target DNA. They applied this DNAzyme-based method to detect the target DNA with a dynamic range from 0.01 to 0.3 μM. Significantly, the two free DNA strands of this DNAzyme could be rationally designed for sensing DNA strands with different sequences. Similar to this research, Kolpashchikov constructed a binary probe to detect the target DNA [40]. He split the peroxidase-like DNA enzyme into two halves and removed the deoxycytidine. The analyte binding arms were added to each half through triethylene glycol linkers.

Traditionally, the DNAzyme is usually split into two equal parts (1:1), but they could easily assemble to form an active aptamer even in the absence of target DNA and then produced a background signal. Zhou's group reported that they split the guanine-rich sequences into 3:1 (one fragment possesses three GGG repeats, and the other possesses one GGG repeat) to solve this problem (Fig. 6.6) [39]. The G-rich segments could integrate into many patterns, thus avoiding forming an active aptamer in the absence of target DNA. To research the probe architectural features and the reaction conditions that affect DNAzyme-based catalytic efficiency, Sintim and his coworkers have performed a series of experiments by using active DNAzyme catalysts for nucleic acid sensing [43]. They showed that the loops that connect G3-tracts in G-quadruplexs structure could be replaced with a stem-loop or loop-stem-loop motif without destabilizing the resulting G-quadruplex structure. In addition, they indicated that the addition of hemin could lead to the conformational

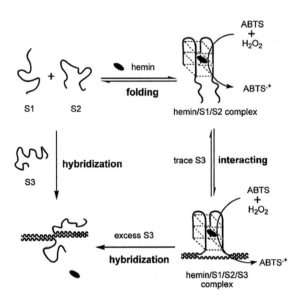

**Fig. 6.5** Scheme of label-free colorimetric method to assay the target DNA based on the supramolecular hemin-G-quarter complex (Reprinted with permission from Ref. [41]. Copyright 2007 The Royal Society of Chemistry). The adjacent positions of target DNA could hybridize with the binary DNA structure, assembling in a G-quadruplex structure. This structure possessed peroxidase activity and could catalyze the $H_2O_2$-mediated oxidation of the substrate to the colored product. This system can distinguish the full matched target and mismatched target well

transition of G-quadruplex from antiparallel to parallel-/mixed-type structures. Thus, they concluded that the stabilization of these structures might result in higher affinity of G-quadruplexs for hemin. For the reaction condition, they demonstrated that compared with the decomposition rate of G-quadruplex peroxidases in the buffer that contains excess ammonium cation, the G-quadruplex peroxidases decomposed faster in the buffer that possesses excess potassium or sodium cations. Significantly, they found that both the sequence content and length of the sequences that flank the G-tracts affected the enzymatic catalysis and background noise. The discovery of these salient architectural designs can facilitate the colorimetric detection of target DNA with picomolar concentrations using label-free DNA probes.

By employing the signal amplification strategy to the colorimetric assay, the detection sensitivity for nucleic acid can be significantly improved [8, 34]. Zhang's group took the advantage of the topological effect of G-quadruplex/hemin DNAzyme, combining DNAzyme assistant DNA recycling and recycling followed with rolling circle amplification strategy to design a colorimetric DNA detection (Fig. 6.7) [45]. They achieved the target DNA detection at a low level (as low as 3.3 fM) with high specificity. Similarly, Ding's group reported a colorimetric strategy for ultrasensitive assay [36]. They combined molecular beacon initiated strand displacement amplification and catalytic hairpin assembly

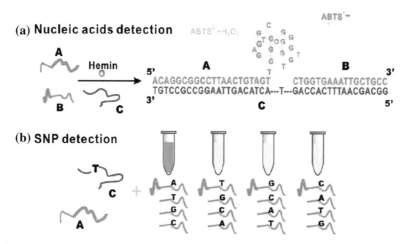

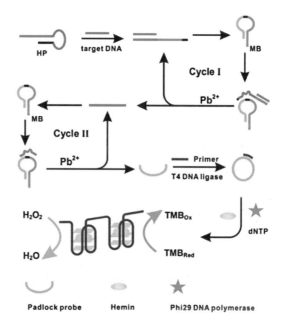

**Fig. 6.6** **a** Schematic illustration of DNA detection using the 3:1 split DNAzyme and **b** principle of single-nucleotide polymorphism detection based on the split DNAzyme (Reprinted with permission from Ref. [39]. Copyright 2008 American Chemical Society)

**Fig. 6.7** Scheme of colorimetric sandwich biosensor for target cancer gene assay based on DNAzyme assistant DNA recycling and recycling followed with rolling circle amplification strategy (Reprinted with permission from Ref. [45]. Copyright 2015 The Royal Society of Chemistry)

(CHA) with DNAzyme formation. The presence of target nucleic acid induced the strand displacement amplification to release nicking DNA triggers, which initiated CHA to produce many CHA products. The CHA products could integrate with hemin to form G-quadruplex/hemin DNAzyme, and then catalyzed a colorimetric reaction. This DNAzyme-based colorimetric assay showed high sensitivity in a dynamic response range from 5 fM to 5 nM.

## 6.5 Conclusion

In this chapter, we have described some of the recent developments in the detection of nucleic acid by colorimetric sandwich assay. The advancement of the colorimetric sandwich assay described here is based on the nanoparticles-, traditional enzyme-, and DNAzyme-based sensor. Some efforts have been strongly made to improve the performance of colorimetric sandwich assay for nucleic acids from all aspects. For example, the employments of AuNPs and enzymes in sandwich assay significantly amplify the signal and improve the detection sensitivity. In addition, the developed asymmetrically PEGylated AuNPs make the assay more stable in complex samples. Thereby, the colorimetric sandwich assay can be viewed as a promising strategy toward detection of nucleic acids, and the full potential of which has yet to be further developed.

## References

1. Rosi NL, Mirkin CA (2005) Nanostructures in biodiagnostics. Chem Rev 105:1547–1562
2. Saha K, Agasti SS, Kim C, Li XN, Rotello VM (2012) Gold nanoparticles in chemical and biological sensing. Chem Rev 112:2739–2779
3. Zhou W, Gao X, Liu DB, Chen XY (2015) Gold nanoparticles for in vitro diagnostics. Chem Rev 115:10575–10636
4. Chen J, Qiu HD, Zhang ML, Gu TN, Shao SJ, Huang Y, Zhao SL (2015) Hairpin assembly-triggered cyclic activation of a DNA machine for label-free and ultrasensitive chemiluminescence detection of DNA. Biosens Bioelectron 68:550–555
5. Lu SS, Hu T, Wang S, Sun J, Yang XR (2017) Ultra-sensitive colorimetric assay system based on the hybridization chain reaction-triggered enzyme cascade amplification. ACS Appl Mater Interfaces 9:167–175
6. Zhao QL, Zhang Z, Xu L, Xia T, Li N, Liu JL, Fang XH (2014) Exonuclease I aided enzyme-linked aptamer assay for small-molecule detection. Anal Bioanal Chem 406:2949–2955
7. Chen XP, Zhou DD, Shen HW, Chen H, Feng WL, Xie GM (2016) A universal probe design for colorimetric detection of single-nucleotide variation with visible readout and high specificity. Sci Rep 6:20257
8. Long YY, Zhou CS, Wang CM, Cai HL, Yin CY, Yang QF, Xiao D (2016) Ultrasensitive visual detection of HIV DNA biomarkers via a multi-amplification nanoplatform. Sci Rep 6:23949
9. Xu H, Wu D, Li CQ, Lu Z, Liao XY, Huang J, Wu ZS (2017) Label-free colorimetric detection of cancer related gene based on two-step amplification of molecular machine. Biosens Bioelectron 90:314–320
10. Xu JG, Qian J, Li HL, Wu ZS, Shen WY, Jia L (2016) Intelligent DNA machine for the ultrasensitive colorimetric detection of nucleic acids. Biosens Bioelectron 75:41–47

11. Song GT, Chen CE, Ren JS, Qu XG (2009) A simple, universal colorimetric assay for endonuclease/methyltransferase activity and inhibition based on an enzyme-responsive nanoparticle system. ACS Nano 3:1183–1189

12. Pinijsuwan S, Shipovskov S, Surareungchai W, Ferapontova EE, Gothelf KV (2011) Development of a lipase-based optical assay for detection of DNA. Org Biomol Chem 9:6352–6356

13. Xie XJ, Xu W, Liu XG (2012) Improving colorimetric assays through protein enzyme-assisted gold nanoparticle amplification. Acc Chem Res 45:1511–1520

14. Cordray MS, Amdahl M, Richards-Kortum RR (2012) Gold nanoparticle aggregation for quantification of oligonucleotides: optimization and increased dynamic range. Anal Biochem 431:99–105

15. Li HB, Wu ZS, Shen ZF, Shen GL, Yu RQ (2014) Architecture based on the integration of intermolecular G-quadruplex structure with sticky-end pairing and colorimetric detection of DNA hybridization. Nanoscale 6:2218–2227

16. Elghanian R, Storhoff JJ, Mucic RC, Letsinger RL, Mirkin CA (1997) Selective colorimetric detection of polynucleotides based on the distance-dependent optical properties of gold nanoparticles. Science 277:1078–1081

17. Taton TA, Mirkin CA, Letsinger RL (2000) Scanometric DNA array detection with nanoparticle probes. Science 289:1757–1760

18. Nguyen HK, Southern EM (2000) Minimising the secondary structure of DNA targets by incorporation of a modified deoxynucleoside: implications for nucleic acid analysis by hybridisation. Nucleic Acids Res 28:3904–3909

19. Ishibashi M, Arakawa T, Philo JS, Sakashita K, Yonezawa Y, Tokunaga H, Tokunaga M (2002) Secondary and quaternary structural transition of the halophilic archaeon nucleoside diphosphate kinase under high- and low-salt conditions. FEMS Microbiol Lett 216:235–241

20. Tan ZJ, Chen SJ (2006) Nucleic acid helix stability: effects of salt concentration, cation valence and size, and chain length. Biophys J 90:1175–1190

21. Menhaj AB, Smith BD, Liu JW (2012) Exploring the thermal stability of DNA-linked gold nanoparticles in ionic liquids and molecular solvents. Chem Sci 3:3216–3220

22. Zu YB, Ting AL, Yi GS, Gao ZQ (2011) Sequence-selective recognition of nucleic acids under extremely low salt conditions using nanoparticle probes. Anal Chem 83:4090–4094

23. Shen W, Deng HM, Gao ZQ (2012) Gold nanoparticle-enabled real-time ligation chain reaction for ultrasensitive detection of DNA. J Am Chem Soc 134:14678–14681

24. Storhoff JJ, Lazarides AA, Mucic RC, Mirkin CA, Letsinger RL, Schatz GC (2000) What controls the optical properties of DNA-linked gold nanoparticle assemblies? J Am Chem Soc 122:4640–4650

25. Thanh NTK, Rosenzweig Z (2002) Development of an aggregation-based immunoassay for anti-protein A using gold nanoparticles. Anal Chem 74:1624–1628

26. Guo LH, Xu Y, Ferhan AR, Chen GN, Kim DH (2013) Oriented gold nanoparticle aggregation for colorimetric sensors with surprisingly high analytical figures of merit. J Am Chem Soc 135:12338–12345

27. Le Goff GC, Blum LJ, Marquette CA (2011) Enhanced colorimetric detection on porous microarrays using in situ substrate production. Anal Chem 83:3610–3615

28. Wu Z, Wu ZK, Tang H, Tang LJ, Jiang JH (2013) Activity-based DNA-gold nanoparticle probe as colorimetric biosensor for DNA methyltransferase/glycosylase assay. Anal Chem 85:4376–4383

29. Garcia J, Zhang Y, Taylor H, Cespedes O, Webb ME, Zhou DJ (2011) Multilayer enzyme-coupled magnetic nanoparticles as efficient, reusable biocatalysts and biosensors. Nanoscale 3:3721–3730

30. Su X, Teh HF, Lieu XH, Gao ZQ (2007) Enzyme-based colorimetric detection of nucleic acids using peptide nucleic acid-immobilized microwell plates. Anal Chem 79:7192–7197

31. Zhang N, Appella DH (2007) Colorimetric detection of anthrax DNA with a peptide nucleic acid sandwich-hybridization assay. J Am Chem Soc 129:8424–8425

32. Li J, Song SP, Liu XF, Wang LH, Pan D, Huang Q, Zhao Y, Fan CH (2008) Enzyme-based multi-component optical nanoprobes for sequence-specific detection of DNA hybridization. Adv Mater 20:497–498
33. Li J, Song SP, Li D, Su Y, Huang Q, Zhao Y, Fan CH (2009) Multi-functional crosslinked Au nanoaggregates for the amplified optical DNA detection. Biosens Bioelectron 24:3311–3315
34. Brown CW, Lakin MR, Horwitz EK, Fanning ML, West HE, Stefanovic D, Graves SW (2014) Signal propagation in multi-layer DNAzyme cascades using structured chimeric substrates. Angew Chem Int Ed 53:7183–7187
35. Willner I, Shlyahovsky B, Zayats M, Willner B (2008) DNAzymes for sensing, nanobiotechnology and logic gate applications. Chem Soc Rev 37:1153–1165
36. Yan YR, Shen B, Wang H, Sun X, Cheng W, Zhao H, Ju HX, Ding SJ (2015) A novel and versatile nanomachine for ultrasensitive and specific detection of microRNAs based on molecular beacon initiated strand displacement amplification coupled with catalytic hairpin assembly with DNAzyme formation. Analyst 140:5469–5474
37. Yang XH, Wang Q, Wang KM, Tan WH, Li HM (2007) Enhanced surface plasmon resonance with the modified catalytic growth of Au nanoparticles. Biosens Bioelectron 22:1106–1110
38. D'Agata R, Corradini R, Grasso G, Marchelli R, Spoto G (2008) Ultrasensitive detection of DNA by PNA and nanoparticle-enhanced surface plasmon resonance imaging. ChemBioChem 9:2067–2070
39. Deng MG, Zhang D, Zhou YY, Zhou X (2008) Highly effective colorimetric and visual detection of nucleic acids using an asymmetrically split peroxidase DNAzyme. J Am Chem Soc 130:13095–13102
40. Kolpashchikov DM (2008) Split DNA enzyme for visual single nucleotide polymorphism typing. J Am Chem Soc 130:2934–2935
41. Li T, Dong SJ, Wang EK (2007) Enhanced catalytic DNAzyme for label-free colorimetric detection of DNA. Chem Commun 4209–4211
42. Li T, Wang E, Dong SJ (2008) Chemiluminescence thrombin aptasensor using high-activity DNAzyme as catalytic label. Chem Commun 5520–5522
43. Nakayama S, Sintim HO (2009) Colorimetric split G-quadruplex probes for nucleic acid sensing: improving reconstituted DNAzyme's catalytic efficiency via probe remodeling. J Am Chem Soc 131:10320–10333
44. Travascio P, Witting PK, Mauk AG, Sen D (2001) The peroxidase activity of a hemin-DNA oligonucleotide complex: free radical damage to specific guanine bases of the DNA. J Am Chem Soc 123:1337–1348
45. Liang D, You W, Yu Y, Geng Y, Lv F, Zhang B (2015) A cascade signal amplification strategy for ultrasensitive colorimetric detection of BRCA1 gene. RSC Adv 5:27571–27575

# Chapter 7
# Fluorescence Sandwich Assays for Nucleic Acid Detection

Xinwen Liu and Quan Yuan

**Abstract** Fluorescence sandwich assays have wide application in the detection of nucleic acids due to its well-developed synthesis process, simple detection procedures, and high sensitivity. Specifically, two oligonucleotide probes, named capture probe and signal probe respectively, are introduced and hybridize with different regions of a single-stranded target gene, forming a "capture probe-target-signal probe" sandwiched format. Distinctive fluorescent emission is therefore generated with the formation of sandwich-format and can be directly detected by conventional instruments without further procedures. In this chapter, we conclude the principle and recent developments of this assay based on the classification of fluorophore materials, including fluorescent organic dyes and fluorescent nanomaterials. For each section, the principle of design strategy is firstly introduced, which contains fluorescence resonance energy transfer (FRET) and DNA hybridization-induced fluorescence enhancement. Furthermore, we discuss the limitations and challenges in the development of fluorescent sandwich assays regarding sensitivity and multiple detection capacity, thus providing an overview of the developing situation and offering insight to further developments of nucleic acid assay.

**Keywords** Fluorescence sandwich assay · Nucleic acid · Organic dye
Quantum dot · Dye-doped nanoparticle

The original version of this chapter was revised: Foreword has been included and authors' affiliations have been updated. The erratum to this chapter is available at https://doi.org/10.1007/978-981-10-7835-4_13

X. Liu · Q. Yuan (✉)
College of Chemistry and Molecular Sciences, Wuhan University,
Wuhan 430072, People's Republic of China
e-mail: yuanquan@whu.edu.cn

X. Liu
e-mail: xinwenliu@whu.edu.cn

## 7.1  Introduction

The obvious advantages of fluorescence-based assay, which include high sensitivity, rapid detection ability, and multiple detection capacity, make it a promising
platform for the analysis of nucleic acid. Traditional fluorescence detection techniques are fluorescent in situ hybridization and *Taqman* real-time PCR detection
[1–3]. However, laborious labeling, elution procedures, or extra amplification steps
are required in such detection process, and the selectivity of analyte is also limited.
Therefore, it is important to develop high-sensitivity, simple, and low-cost nucleic
acid detection methods. Fluorescence sandwich assay has attracted increasing
attention due to its simple design principle, well-developed synthesis methods, and
confined detection procedures. Generally, two oligonucleotide probes, named
capture probe and signal probe (with fluorophore labeling), are introduced for
detection. These two probes hybridize with different regions of a target oligonucleotide sequence, forming a capture-probe-target-signal-probe-sandwiched format.
Distinctive fluorescent emission is therefore generated and can be directly detected
by conventional instruments without further procedures. Analogous to other
fluorescence detection methods, the development of this assay is accompanied with
the invention of fluorophores with better properties, the application of different
optical effects, and the innovation of detection apparatus, which will be specifically
discussed as following.

## 7.2  Organic Dyes as Fluorophores for Sandwich Assays

Organic dyes have significant application in fluorescence detection due to their
availability from commercial sources, small size, and compatibility with various
covalent coupling strategies [4, 5]. Moreover, for nucleic acid analysis, organic
dyes can be attached at the terminus or any internal position of nucleic acid through
suitable linker arms [4]. The well-developed conjugation methods greatly benefit
the design of fluorescence sandwich assay for nucleic acids' detection.

In this assay, the strategies that employ organic dye as fluorophore can be
divided into two main categories, one is with labeled dyes and the other is the
label-free method [4]. The former strategy usually exploits fluorescence resonance
energy transfer (FRET) or excimer formation optical effect; while in the latter
strategy, intercalation method is frequently used.

## 7.2.1 Sandwich Assays Based on Fluorescence Resonance Energy Transfer (FRET)

FRET is a non-radiative process contains energy transfer between an excited state donor and a ground state acceptor [6]. As this process stems from dipole–dipole interactions, it is extremely dependent on the separation distance between donor and acceptor ($R$). Moreover, it also requires a suitable orientation relationship of donor and acceptor, and an appropriate spectral overlap between donor emission and acceptor absorption [5]. The transfer efficiency of FRET is proportional to the inverse sixth power of the molecule distance ($R^{-6}$), the twice power of orientation factor ($\kappa^2$), and also linearly proportional with the spectral overlap integral ($J$) [6]. This can be expressed in the following equation, where $R_0$ is the Förster distance, $n$ is the medium refraction index, $f_D(\lambda)$ is the donor emission spectrum, and $\varepsilon_A(\lambda)$ is the acceptor absorption spectrum.

$$E = \frac{1}{1 + \left(\frac{R}{R_0}\right)^6} \tag{7.1.1}$$

$$R_0^6 = \frac{9000(\ln 10)\kappa^2 \phi J}{128\pi^5 n^4 N_{AV}} \tag{7.1.2}$$

$$J = \int_0^\infty f_D(\lambda)\varepsilon_A(\lambda)\lambda^4 d\lambda \tag{7.1.3}$$

In particular, for nucleic acid sandwich assay, two oligonucleotide probes, which are complementary to adjacent regions of nucleic acid target, are modified with a donor and an acceptor at the 3' or 5' terminus, respectively. Without the target, because of the low concentration of probes, the distance between randomly diffused acceptors and donors is not close enough for FRET. Therefore, the donor emission is unaffected by acceptors. Upon the target binding, the two probes hybridize with adjacent positions of the target, which brings them in close proximity for FRET. When exciting the donor groups, energy is transferred from donors to acceptors, thus decreasing the signal from donors. This model is also referred to as binary or adjacent probes in DNA hybridization, first reported by Heller and Morrison [7]. The donor should be fluorophore to provide excitation energy in FRET, but the acceptor can be either quencher or fluorophore. Quenchers are not fluorescent and therefore cause the fluorescence of donors simply to decrease [3]. In other words, it is a "signal-off" assay, whose limitation is that the signal change can be over 100% (Fig. 7.1, left) [8]. On the other hand, when the acceptor is fluorescent, it absorbs the energy from donor, which results in the decrease of donor fluorescence and the increase of acceptor fluorescence, making it a "signal-on" assay (Fig. 7.1, right). As two measureable parameters are created in the latter assay, it has wider application compared with the signal-off assay.

**Fig. 7.1** Sandwich assay for
nucleic acids based on FRET
(left, middle) and an
intercalation dye (right).
(Reprinted with permission
from Ref. [2]. Copyright 2014
American Chemical Society)

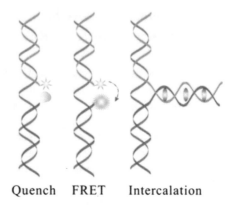

Quench    FRET    Intercalation

Organic dyes are the first generation of donor–acceptor pairs in FRET applications, tested with novel fluorescent materials [4, 5]. For nucleic acid sandwich
assays, typical donor fluorophores are cyanine dye (Cy3) and 6-carboxyfluorescein
(6-FAM); the acceptor fluorophores include the cyanine dye (Cy5), 6-carboxy-N,N,
N′,N′-tetramethylrhodamine (TAMRA), and Bodipy493/503 [9]. To maximize the
transfer efficiency, the distance between acceptor and donor is recommended with
1–5 bases. Too close distance might lead to direct interaction between two fluorophores, decreasing the signal intensity [10]. Suitable spectrum overlap between
acceptor and donor fluorophores is also required. These are important factors that
need to be taken into account of in the design of each probe. As for application, the
organic   dye-based   sandwich   assay   has   been   used   to   detect   quantity,
single-nucleotide polymorphism, fragment secondary structure, gene translocation
process, and imaging observation of nucleic acid target [10–14]. For instance, Tsuji
et al. employed this assay for the observation of messenger RNA imaging in living
cells [10]. Bodipy493/503 and Cy5 were used as donor and acceptor fluorophores,
respectively. When these two probes hybridized with target c-fos mRNA, sandwich
format formed and their close proximity resulted in efficient FRET, leading to the
decrease of donor fluorescence (503 nm) and the increase of acceptor fluorescence
(664 nm). As the efficient acceptor fluorescence signal only occurred in the presence of target, it was used for target localization under the observation of fluorescent microscopy. After injecting streptavidin-modified donor and acceptor probes to
living Cos7 cells, successful imaging and localization of c-fos mRNA in single cell
were achieved.

Despite the widespread application of this sandwich assay, the main problem is its
rather high background noise, caused by the direct excitation of non-hybridized
acceptor probes. Although it is a possible solution to reduce the spectral overlap
between the donor and acceptor, the intensity of the FRET signal also correspond-
ingly decreases, as shown in Eqs. 7.1.1 and 7.1.2. Three-dye binary probes have
been invented to improve the signal-to-background (S/B) ratio of this assay, which is
constructed by inserting two fluorophores into the donor probe [6, 15, 16]. The
inserted two fluorophores can be either identical [16] or not [6]. Still employing the

Bodipy493/503-Cy5 pairs, Watanabe et al. labeled two Bodipy493/503 fluorophores on the donor probe. Compared with the ordinary probes, this three-dye system produced a considerable increase in acceptor emission because of the enhanced transfer energy. Moreover, the background noise of this system was also weaker than that of the single-labeled one, possibly due to increased self-quenching of the acceptor fluorophore. Therefore, the S/B ratio was increased [16]. Turro et al. created a FAM–TAMRA–Cy5 FRET system, in which a FAM fluorophore and a TAMRA fluorophore were labeled on the same donor probe and separated by four nucleotides. In the absence of target, because of the close proximity of FAM and TAMRA, strongest fluorescence of intermediate TAMRA could be detected. With the target, energy was transferred from the donor FAM through the intermediate fluorophore TAMRA to the acceptor Cy5, leading to the most intense emission from the acceptor Cy5. Compared with the three-dye system employing identical donor fluorophores, the introducion of an intermediate fluorophore (TAMRA) makes it possible to choose the donor and acceptor with even smaller spectral overlap, reducing the direct excitation and also providing efficient energy reaching acceptor [6].

Another approach is to employ molecular beacons (MBs) in the construction of either donor probe [17] or both the donor and acceptor probes [18]. MBs are dual-labeled hairpin-shaped oligonucleotide probes with a reporter fluorophore at one end and a quencher at the other end. Without the target, the reporter and quencher are brought in close proximity by the stem formation, resulting in the FRET-based quenching of the reporter and thus limiting the background signals. However, in the presence of the target, the probe hybrids with the target and undergoes a conformational reorganization, which disrupts the FRET and restores the fluorescence of donor [17, 18]. When introducing MBs to the sandwich assay, the background signals, caused by current emission of unhybridized donors and direct excitation of acceptors, can be effectively reduced by the intramolecular quenching effect of MBs [9]. Bao et al. designed a sandwich assay employing two MB probes as donor and acceptor, respectively [18]. The donor MB probe was attached with a BHQ-2 quencher to the 3′ end and a Cy3 fluorophore to the 5′ end. The acceptor MB probe was labeled with a BHQ-3 quencher to the 5′ end and a Cy5 fluorophore to the 3′ end (Fig. 7.2). When both the MB probes hybridize with target mRNA and thus undergo conformational change, FRET happens between the Cy3 fluorophore and Cy5 fluorophore. Employing these dual MB probes, localization of the K-ras mRNA expression in living HDF cells was achieved with rather low background interference.

## 7.2.2  Nucleic Acid Sandwich Assay Based on DNA Hybridization-Induced Fluorescence Enhancement

An excimer is a dimer that is formed by two molecules of the same species. One molecule is in the excited state, and the other is in the ground state. The simplest

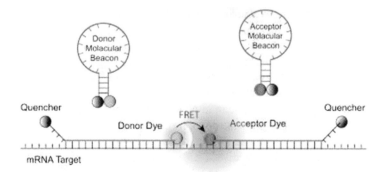

**Fig. 7.2** Scheme of dual molecular beacon (MBs)-based nucleic acid sandwich assay. Two MBs function as donor probe and acceptor probe, respectively. In the presence of targets, both the probes undergo conformational reorganization and hybridize with adjacent parts of the target, allowing for FRET between donor dye and acceptor dye. (Reprinted with permission from Ref. [18]. Copyright 2004 Oxford University Press)

formation of an excimer can be shown in equation below ($^1M^*$ is a singlet excited molecule, $^1M$ is an unexcited molecule, and $^1D^*$ is the excited dimer) [19]. The excited monomer will relax to the steady singlet $S^1$ state through internal conversion at first and then constitute a lower energy excimer when meeting with a ground state molecule. In the later dissociation process, the lower energy excimer will release a photon at a longer wavelength than the excited monomer does. Because of the characteristic longer wavelength region and broad band of excimer emission, it can be easily distinguished from the monomer fluorescence, allowing for the application of excimer-based probes in real-time hybridization assays [19, 20].

$$^1M^* + {}^1M \leftrightharpoons {}^1D^* \tag{7.2.1}$$

$$^1D^* \rightarrow {}^1M + {}^1M + h\nu_D \tag{7.2.2}$$

Specifically, when applying excimer formation principle to sandwich assay, the main construction of the sandwiched structure is similar with that of the FRET-based assay. The only difference lies in the choice of labels on probes. Both strands of excimer probes are attached with a monomer molecule which is able to form excimer in close proximity. This assay was also referred to as excimer-forming two-probe hybridization method, firstly reported by Ebata and his coworkers [20]. Pyrene, the well-characterized monomer, was chosen as a fluorophore and was attached on two separated oligonucleotide probes (Fig. 7.3). When these probes hybridized to the target Vibrio mimicus 16S rRNA, a 495-nm broad fluorescence band spectra were easily discriminated from the monomer fluorescence (below 450 nm). With the growing concentration of targets, the intensity of this band increased and that of the pyrene monomer decreased, indicating that the 495-nm band was attributed to the pyrene excimer. A detection limit of 10 nM was achieved in this system [21].

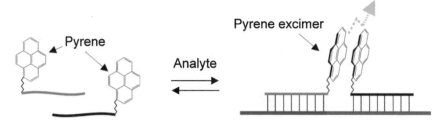

**Fig. 7.3** Scheme of dual excimer formation-based nucleic acid sandwich assay. Two target-complementary probes are modified with a pyrene monomer. When hybridizing with target, the close proximity between the two monomer allows the formation of an excimer, generating distinctive fluorescence signal. (Reprinted with permission from Ref. [9]. Copyright 2010 American Chemical Society)

The main advantage of this assay is that excimer's large Stoke shift allows for the use of a wider spectral band-pass filters for the detection channels, which is beneficial to signal intensity enhancement [22]. However, to avoid the quenching effect caused by nearby nucleic acids through photo-induced electron-transfer (PET) process, 20% (v/v) DMF has to be added in the hybridization buffer to provide efficient signal intensity [21], which is not suitable for in vivo applications. Moreover, the excitation wavelength of monomer is at around 340 nm, which will cause limited depth penetration and serious auto fluorescence of cellular environment. Fortunately, another feature of excimers can also be applied to cell studies. Their relatively long lifetime (30–70 ns), in comparison with the short lifetime of the auto fluorescence background in biological sample (8 ns), enables the use of time-resolved fluorescence spectroscopy for signal discrimination [19, 22]. Turro et al. employed this method for mRNA detection in neuronal extracts [22]. Although the cellular extracts showed rather high background noise after 340 nm excitation, the emission decay of this noise was 8 times faster than that of the pyrene excimer probes. The pyrene excimer emission signal was thus gated in the interval of 30–150 ns, giving an S/B ratio of $\sim 10$.

The aforementioned approaches all involve dye labeling in the construction of oligonucleotide probes, which requires extra chemical synthesis and purify steps and thus improves the cost. To avoid chemical modifications, label-free sandwich methods have been invented with the development of organic dye-specific aptamers. Aptamers are artificial functional oligonucleotides that are able to bind to specific target molecules. They are selected from large random DNA or RNA pools utilizing in vitro selection (SELEX) approach [9]. During the synthesis, only target-binding molecules are isolated as aptamers so that they have great combination ability with targets. They are also known as the strong rival to antibody. Aptamers have been selected toward a broad range of targets, including organic dyes. For instance, malachite green (MG)-binding aptamers were first reported in 1998 [23], which increased the dye fluorescence more than 2000-fold upon binding by stabilizing MG in a planar (more fluorescent) conformation [24]. Kolpashchikov firstly used this aptamer for sequence detection of nucleic acids [25]. In this work, the RNA aptamer was separated into two strands. Both strands were attached with target-complementary nucleic acid binding arms through UU dinucleotide bridges (Fig. 7.4). Without the DNA target, MG and probes were separated and the fluorescence of MG was comparably weak. In the presence of the target, both strands hybridized with adjacent potions of target and reform the MG aptamer, constructing a non-conical sandwich structure complex. Strong fluorescence increase ($\sim 20$ times higher) was therefore produced due to the enhancement effect of MG aptamers. The detection limit of this system was around of 100 nM.

The development of this assay is accompanied with the invention of aptamers with larger enhancement effect and stronger binding ability. After the production of MG aptamer, Hoechst dye-aptamer [26, 27] has also been synthesized for single-nucleotide resolution, with nearly 143 times fluorescence enhancement compared with direct dye excitation. Recently, an RNA molecule named "Spinach aptamer" with affinity for a fluorescent dye DFHBI was produced, which has great enhancement effect with nearly negligible photobleaching [28]. Exploiting this Spinach system, successful RNA recognition was achieved, which yielded a 270 times higher fluorescence after binding and obtained a LOD at 1.8 nM [29], superior to the existed intercalation-based system.

Conjugated polymers (CPs) are organic macromolecules comprising at least one chain of alternating double bond and single bond. Their delocalized $\pi$ electron system allows for light absorption and photo-generated charge carrier generation, making them the ideal light-harvesting material [30]. Moreover, the excitation can migrate along the polymer chain through efficient energy transfer. Thus, CPs are promising candidates for FRET donor materials [31].

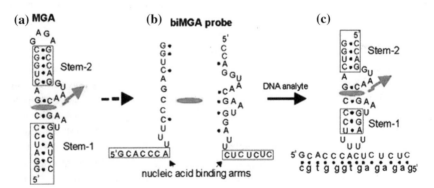

**Fig. 7.4** Scheme of intercalation-based nucleic acid sandwich assay (MG aptamer). **a** The structure of MG aptamer. **b** The MG aptamer is separated into two strands, both of which are attached with target-complementary nucleic acid binding arms. **c** In the presence of DNA analyte, the two strands hybridize with target, reforming the MG aptamer. High S/B ratio can be obtained due to the fluorescence amplification effect of MG aptamer. (Reprinted with permission from Ref. [25]. Copyright 2005 American Chemical Society)

In particular, for DNA sandwich assay, a cationic CP functions as the capture probe, and a target-complementary PNA strand labeled with a fluorescent dye acts as the acceptor probe. In the presence of target, the PNA probe forms stable Watson–Crick base pairs with the single-stranded DNA target. The cationic CP then combines with the polyanionic PNA/DNA complex through electrostatic interaction, constructing a probe-target-probe-sandwiched format (Fig. 7.5) [32]. The close proximity of CP and the fluorophore on PNA enables FRET and therefore generating fluorescent signals of the fluorophore. In PNA, the negatively charged phosphate linkages in conventional DNA are replaced with peptomimetic neutral amide linkages, making it a neutral material. Thus, without the DNA target, no electrostatic interaction happens between cationic CP and neutral PNA probe, limiting the background signal.

Bazan et al. firstly reported this system and employed it for quantification detection of target DNA. A 25-times amplification of fluorescein emission compared with direct dye excitation was achieved, with a detection limit of 10 pM [32]. However, to avoid detrimental background noise caused by the hydrophobic interaction of PNA and cationic CP, 10% ethanol solution was added. In order to further improve the sensitivity of this scheme, label-free methods employing positively charged intercalation dye have been invented [33]. Specifically, cationic CP (PFP), target-complementary single-strand DNA (ssDNA) probe and intercalation dye, Genefinder (GF), were mixed in the solution with suitable ratio. When adding

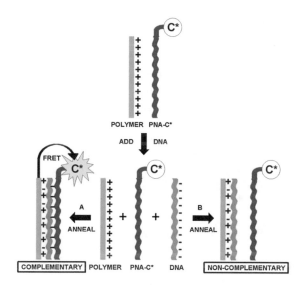

**Fig. 7.5** Scheme of conjugated polymer-assisted sandwich assay. Cationic conjugated polymer functions as donor probe, and fluorophore-labeled PNA functions as acceptor probe. With the ssDNA target, target-complementary PNA forms stable Watson–Crick base pairs with the target. Cationic conjugated polymer combines with the polyanionic PNA/DNA complex through electrostatic interaction. The close proximity between polymer and fluorophore on PNA allows for FRET. (Reprinted with permission from Ref. [32]. Copyright 2002 National Academy of Sciences)

the target, PFP and helix DNA formed a sandwich structure and GF dye intercalated into the helix DNA, which brought GF and PFP in close proximity and thus allowed for FRET. Because both PFP and GF are positively charged, electrostatic repulsion between them should be expected without the target, which effectively reduces the background noise and improves its sensitivity. Another label-free method employing pyrene-functionalized polymer and intercalation dye SYBR Green I has also been reported for selectivity promotion [34].

## 7.3 Nanomaterials as Fluorophores for Sandwich Assays

Although organic dyes have been widely used in fluorescence sandwich assay, they also have certain limitations, such as their self-quenching at high concentrations, susceptibility to photobleaching, and narrow absorption windows with small Stoke shifts, which impede the sensitivity improvement [4, 5].

In comparison with organic dyes, nanomaterial has been developed as the new generation fluorescence material with great optical properties. Specifically, for nucleic acid sandwich assays, quantum dots (QDs) and dye-doped nanoparticles have been applied as fluorophores.

### 7.3.1 Quantum Dots as Fluorophores

QDs are semiconductor nanocrystals with physical dimensions smaller than the exciton Bohr radius, which have great electronic and optical properties [35]. The advantages of QDs, such as strong fluorescence, high photostability, size-tunable narrow emission, and broad excitation window with large Stoke shifts, make them the most promising fluorescent materials.

In particular, for nucleic acid sandwich assay, the strategies for employing QDs are similar with those of the organic dye-based sandwich assay. They can be mainly divided into two categories, which are using QD as the signal output (Fig. 7.6, left) or as the donor in FRET-based assay (Fig. 7.6, right) [2].

#### 7.3.1.1 Quantum Dots as the Signal Output

In this assay, both the target-complementary capture probe and the report probe are labeled with fluorophores that have distinct emission, exploiting the principle of dual-color fluorescence coincidence detection [36]. These two distinct emissions are used to "encode" the target, since only the sandwiched structure reflects the fluorescence of both the two channels and thus is able to be distinguished from backgrounds. Specifically, for QD-based sandwich assay, the luminescence of QDs functions as the reporting signal, analogous to the barcode in code system. QDs

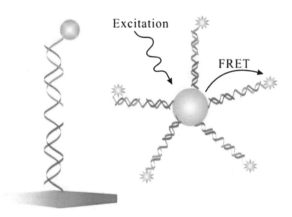

with surface-functionalized oligonucleotide probes can be used as either report
probe [37], capture probe [38], or both in this system [39].

Recently, Kim et al. developed the bead-based sandwich hybridization for rapid
analysis of the Bacillus spoOA gene [37]. Fluorescent bead-DNA complex func-
tioned as the capture probe and QD 655-DNA, constructed via biotin–streptavidin
interaction, was used as the signal probe (Fig. 7.7). Analyzed by flow cytometry,
the area of sandwich complex was determined by four parameters, the scan scatter,
electronic volume, green fluorescence from bead, and red fluorescence from QDs.
This system demonstrated both wider linear range (3.2–1000 nM) and lower
detection limit (0.02 nM) than conventional DNA biosensors.

QD-DNA probe can also act as the capture probe, in which QDs are the
nanoconcentrators of target that have signal amplification effect [38]. Wang et al.
used two ssDNA probes for detection, one modified with biotin and the other
labeled with an organic fluorophore, Oregon Green 488 (peak emission at 524 nm).
After they form a sandwich complex with target, streptavidin-modified QDs (core–
shell CdSe-ZnS, QD 605) were added to capture these biotinylated sandwich
structures. Because a single quantum dot was able to combine with several copies
of sandwiched target, the signal of organic fluorophore was amplified. Confocal
fluorescent spectroscopy was used for target detection, as the coincidence of green
channel and red channel only occurred in the presence of targets. This
nanobiosensors showed better sensitivity and specificity compared with MB-based
assays due to its amplification effect. However, the limitations of organic dyes, such
as narrow absorption windows with small Stoke shifts, result in emission spectral
overlap (also known as cross talk) between organic dyes and QDs, in addition with
the confined choice of excitation wavelength, hindering the next development of
multiple targets detection.

Considering the disadvantage brought about by organic fluorophores, Wang
et al. firstly employed two QD-DNA probes in the colocalization analysis [39].
These probes have discernible emission wavelengths and hybridized with the target
DNA in juxtaposition, forming a sandwiched structure (Fig. 7.8). Analyzed by

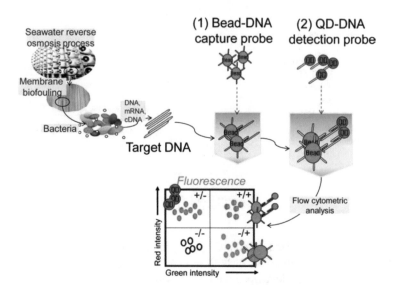

**Fig. 7.7** Scheme for employing quantum dots as reporter probe in sandwich assay. Bead-DNA functions as capture probe, and QD-DNA functions as signal probe. Sandwich complex is formed when these two probes hybridize with targets. Only the sandwich complex reflects both green fluorescence from bead and red fluorescence from QDs. (Reprinted with permission from Ref. [37]. Copyright 2010 American Chemical Society)

confocal fluorescent microscopy, only the target showed emission of two channels (presented by pseudocombined color). Exploiting this assay, simultaneous detection of three genes associated with anthrax pathogenicity was achieved at a single molecule level. Moreover, due to the outstanding optical properties of QDs, such as the broad excitation window and size-tunable emission wavelength, the multiple detection capacity of this assay can be further increased by using extra QDs with distinct emissions. Specifically, $n$ different QDs can simultaneously detect $\frac{1}{2}n(n-1)$ targets.

### 7.3.1.2 Quantum Dots as the Donor in FRET-Based Assays

QDs are promising as FRET donors due to their size-tuned emission and broad absorption window, which avoid spectral cross talk and decrease the direct excitation of acceptors. Also because of their broad absorption width property, it can be inferred that they are not suitable as acceptors. To complete the probe construction, organic dye is frequently used as the acceptor fluorophore in QD-based sandwich assay. Wang et al. firstly employed 605 QD-Cy5 in FRET system [40]. The construction process of sandwich nanoassembly is similar with the group's aforementioned work employing QDs as the nanoconcentrators [38]. The difference lies in whether FRET is involved in the detection. In this 605 QD-Cy5 system, the

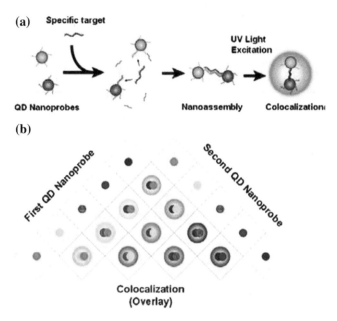

**Fig. 7.8** Scheme for employing two QD-DNA probes in a sandwich assay. **a** Scheme model: QDs with distinctive emission wavelengths are introduced to encode the target. **b** The color combination scheme for multiplexed colocalization detection. (Reprinted with permission from Ref. [39]. Copyright 2005 American Chemical Society)

spectral overlap and close proximity between these two fluorophores allow for FRET (Fig. 7.9). Compared with the conventional organic fluorophore system, the direct excitation of Cy5 can be greatly reduced due to the broad excitation window of QDs, which allows the choice of sample excitation wavelength at near the minimum absorption wavelength of acceptors. Besides, the capability of capturing several acceptors by a single quantum dot improves the overall energy-transfer efficiency. This system showed ∼ 100-fold higher sensing signal than that of MBs and achieved a detection limit of 4.8 fM. For further developments, solid-phase immobilization of different QDs with distinct emissions on optical fibers was introduced in this system [41, 42], which enables multiple target detection. Microfluidic chip with separated channels [43] and paper-based analysis [44] were also invented for detection device construction.

Despite the advantages of QDs as FRET donors, the large size of QDs, contributed by the core–shell CdSe-ZnS structure, extra coating, and bioconjugation, impedes the FRET transfer efficiency, which is strongly dependent on the distance R [10]. Considering the "tail-to-tail" configuration in QD-based sandwich assay, only short-length oligonucleotide probes can be used to ensure the transfer efficiency, which may limit further development. To improve the sensing distance range, QD-gold system was invented, in which nanometer gold particles (AuNPs) quench the fluorescence of QDs with high efficiency in larger distance (up to 15–

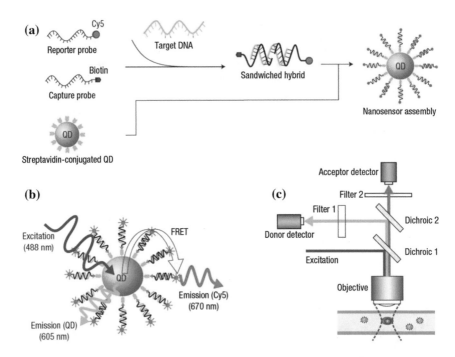

**Fig. 7.9** Scheme for quantum dots as donors in FRET-based sandwich assay. **a** Conceptual scheme of sandwich hybrid formation. QDs function as both the signal probe and target concentrator. **b** Scheme for FRET process between QDs and Cy5 fluorophore. **c** Detection apparatus. (Reprinted with permission from Ref. [40]. Copyright 2005 Nature Publishing Group)

20 nm). Although the quenching mechanism has not been completely proved yet, experiments showed that energy transfer from QDs to small size AuNPs (3 nm) had a nanometal surface energy transfer (NSET)-like $d^{-4}$ dependence, which makes larger distance detection available [45]. Lee et al. constructed a head-to-head sandwiched format in which QD functions as donor and AuNP (5 nm) acts as acceptor. Because of the head-to-head configuration and the possible NSET principle between QDs and AuNPs, oligonucleotide probes with a longer length (30-mer) than the conventional FRET-based detection ($\sim$ 15-mer) could be chosen. Simultaneous detection of two variation types of EML4–ALK fusion genes was achieved by employing green-emitting and red-emitting quantum dots. The detection limit was of 3.45 nM [46].

## 7.3.2 Dye-Doped Nanoparticles as Fluorophore

In addition with quantum dots, to overcome the aforementioned limitations of organic dye fluorophores, dye-doped nanoparticles have also been greatly investigated. Nanoparticles function as the "container" of organic fluorophores, which either encapsulate the fluorophore inside them or attach it on their surface [47]. Among different nanoparticles, silica nanoparticle (SiNP) is the ideal carrier and its dye-nanoparticle assembly is frequently used in DNA sandwich assay.

Compared with conventional organic dyes, dye-dope SiNPs have three main advantages. Firstly, as SiNPs encapsulate a large number of dye molecules inside the silica matrix by absorption, the fluorescence signal is greatly amplified, making it possible for ultrasensitive detection. Secondly, the shielding effect of SiNPs increases the stability of organic dyes. Moreover, the flexible silica chemistry provides versatile routes for surface modification, enabling various biomolecule conjugations [48, 49].

Specifically, for conventional DNA sandwich assays, one most important limitation is that a DNA probe can only be labeled with one or a few fluorophores, prohibiting the ultratrace detection of nucleic acids. One solution is to use dye-doped SiNPs with amplification effect. Zhao and his coworkers developed a sandwich assay based on dye-doped SiNPs [48]. In this work, the biotinylated capture probe (DNA 1) was firstly immobilized on an avidin-coated glass substrate. The signal probe (DNA 3) was labeled with a TMR fluorophore-doped SiNP for signal generation. Both these two probes were complementary to the target. When adding the target (DNA 2) and the capture probes in sequence, the capture probe and the signal probe hybridized with different regions of the target, forming a sandwich structure (Fig. 7.10). Because each SiNP was doped with a large number of fluorophore molecules, the generated signal for even one DNA was greatly amplified. Control experiment proved that the SiNP provided an approximately $10^4$ times higher signal than that of the simple TMR fluorophore in the same assay. By monitoring the fluorescence image and fluorescence intensity, a detection limit as low as 0.8 fM was achieved.

Although dye-doped SiNPs have obvious advantages, several limitations still exist. Because the organic dye molecules are randomly embedded in the silica matrix, fluorophore aggregation occurs, which lead to the decrease of fluorescent efficiency and variation of the amount of dye molecules encapsulated in each

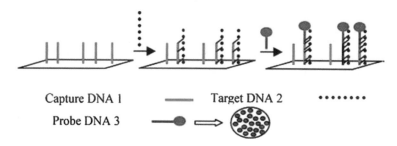

Capture DNA 1        ——        Target DNA 2        ● ● ● ● ● ● ●

Probe DNA 3        ——●  ⇒

**Fig. 7.10** Scheme for dye-doped nanoparticles as fluorophore in a sandwich assay. Biotinylated capture probe is immobilized on the glass substrate; single probe is modified with dye-doped SiNPs. Fluorescence is greatly amplified due to the high fluorophore loading capacity of SiNPs. (Reprinted with permission from Ref. [48]. Copyright 2003 American Chemical Society)

nanoparticle. To prevent such problems, Zhou et al. further developed this assay by creating cyanine dye-doped Au/silica core–shell nanoparticles [50]. The cyanine dyes were firstly conjugated with Au colloidal core through $(dT)_{20}$ oligomers and then embedded in the silica shell, which effectively inhibited the dye blinking problem. The amount of organic fluorophore for each silica shell was also proved to be around of 95 molecules, leading to good reproducibility results. By using Cy3- and Cy5-doped Au/Si NPs in sandwich assay, two-color DNA microarray-based detection was demonstrated and detection limits of 1 pM were obtained.

## 7.4   Conclusion

In this chapter, we summarized the principle of design approaches, the limitations, and related developments of fluorescence sandwich assay. Great efforts have been made for sensitivity improvement, detection capacity enlargement, and application field expansion. The modification of the oligonucleotide probe, which includes inserting two fluorophores into the capture probe and employing MB as probes, has effectively improved the S/B ratio of FRET-based sandwich assay. The exploiting of different design approaches, such as excimer formation and intercalation, either improves the imaging resolution or simplifies the synthesis procedures. The advent of fluorophores with superior optical properties, including CPs, QDs, and dye-doped nanoparticles, greatly promotes its quantification detection. As low as of 0.8 fM detection limit has been obtained with the application of dye-doped SiNPs. Even more sophisticated detection instruments and techniques offer more opportunities for in vivo studies and multiple target analysis. The utilization of time-resolved fluorescence spectroscopy efficiently decreased the autofluorescence signal of in vivo applications. Fluorescence microscopy achieved the imaging and localization of target in biological samples. As for simultaneous detection of multiple targets, the application of confocal fluorescence microscopy and flow cytometry increased precision level of multiple capacities through multiple parameters determination. For further development of this assay, the synthesis of new fluorescent materials and invention of high-throughput detection techniques are still required. By combining the strategies discussed above, such as exploiting design approaches with appropriate probe modifications, and the use of high property fluorophores with state-of-the-art detection techniques, the fluorescence nucleic acid sandwich assay can be further developed.

## References

1. Zuo XB, Yang XH, Wang KM, Tan WH, Wen JH (2007) A novel sandwich assay with molecular beacon as report probe for nucleic acids detection on one-dimensional microfluidic beads array. Anal Chim Acta 587:9–13

2. Shen JW, Li YB, Gu HS, Xia F, Zuo XL (2014) Recent development of sandwich assay based on the nanobiotechnologies for proteins, nucleic acids, small molecules, and ions. Chem Rev 114:7631–7677

3. Epstein JR, Biran I, Walt DR (2002) Fluorescence-based nucleic acid detection and microarrays. Anal Chim Acta 469:3–36

4. Sapsford KE, Berti L, Medintz IL (2006) Materials for fluorescence resonance energy transfer analysis: Beyond traditional donor-acceptor combinations. Angew Chem Int Ed 45:4562–4588

5. Clapp AR, Medintz IL, Mauro JM, Fisher BR, Bawendi MG, Mattoussi H (2004) Fluorescence resonance energy transfer between quantum dot donors and dye-labeled protein acceptors. J Am Chem Soc 126:301–310

6. Marti AA, Li XX, Jockusch S, Stevens N, Li ZM, Raveendra B, Kalachikov S, Morozova I, Russo JJ, Akins DL, Ju JY, Turro NJ (2007) Design and characterization of two-dye and three-dye binary fluorescent probes for mRNA detection. Tetrahedron 63:3591–3600

7. Heller M, Morrison L, Prevatt W, Akin C (1983) Light-emitting polynucleotide hybridization diagnostic method. European patent application 70:685

8. Lubin AA, Plaxco KW (2010) Folding-based electrochemical biosensors: the case for responsive nucleic acid architectures. Acc Chem Res 43:496–505

9. Kolpashchikov DM (2010) Binary probes for nucleic acid analysis. Chem Rev 110:4709–4723

10. Tsuji A, Koshimoto H, Sato Y, Hirano M, Sei-Iida Y, Kondo S, Ishibashi K (2000) Direct observation of specific messenger RNA in a single living cell under a fluorescence microscope. Biophys J 78:3260–3274

11. Mergny JL, Boutorine AS, Garestier T, Belloc F, Rougee M, Bulychev NV, Koshkin AA, Bourson J, Lebedev AV, Valeur B, Thuong NT, Helene C (1994) Fluorescence energy-transfer as a probe for nucleic-acid structures and sequences. Nucleic Acids Res 22:920–928

12. Didenko VV (2001) DNA probes using fluorescence resonance energy transfer (FRET): designs and applications. Biotechniques 31:1106–1107

13. Cardullo RA, Agrawal S, Flores C, Zamecnik PC, Wolf DE (1988) Detection of nucleic-acid hybridization by nonradiative fluorescence resonance energy-transfer. Proc Natl Acad Sci USA 85:8790–8794

14. Masuko M, Ohuchi S, Sode K, Ohtani H, Shimadzu A (2000) Fluorescence resonance energy transfer from pyrene to perylene labels for nucleic acid hybridization assays under homogeneous solution conditions. Nucleic Acids Res 28:E34

15. Juskowiak B (2011) Nucleic acid-based fluorescent probes and their analytical potential. Anal Bioanal Chem 399:3157–3176

16. Okamura Y, Kondo S, Sase I, Suga T, Mise K, Furusawa I, Kawakami S, Watanabe Y (2000) Double-labeled donor probe can enhance the signal of fluorescence resonance energy transfer (FRET) in detection of nucleic acid hybridization. Nucleic Acids Res 28:E107

17. Root DD, Vaccaro C, Zhang ZL, Castro M (2004) Detection of single nucleotide variations by a hybridization proximity assay based on molecular beacons and luminescence resonance energy transfer. Biopolymers 75:60–70

18. Santangelo PJ, Nix B, Tsourkas A, Bao G (2004) Dual FRET molecular beacons for mRNA detection in living cells. Nucleic Acids Res 32:E57

19. Birks J (1975) Excimers. Rep Prog Phys 38:903–974

20. Ebata K, Masuko M, Ohtani H, Kashiwasakejibu M (1995) Nucleic-acid hybridization accompanied with excimer formation from 2 pyrene-labeled probes. Photochem Photobiol 62:836–839

21. Masuko M, Ohtani H, Ebata K, Shimadzu A (1998) Optimization of excimer-forming two-probe nucleic acid hybridization method with pyrene as a fluorophore. Nucleic Acids Res 26:5409–5416

22. Marti AA, Li XX, Jockusch S, Li ZM, Raveendra B, Kalachikov S, Russo JJ, Morozova I, Puthanveettil SV, Ju JY, Turro NJ (2006) Pyrene binary probes for unambiguous detection of mRNA using time-resolved fluorescence spectroscopy. Nucleic Acids Res 34:3161–3168
23. Grate D, Wilson C (1999) Laser-mediated, site-specific inactivation of RNA transcripts. Proc Natl Acad Sci USA 96:6131–6136
24. Babendure JR, Adams SR, Tsien RY (2003) Aptamers switch on fluorescence of triphenylmethane dyes. J Am Chem Soc 125:14716–14717
25. Kolpashchikov DM (2005) Binary malachite green aptamer for fluorescent detection of nucleic acids. J Am Chem Soc 127:12442–12443
26. Sando S, Narita A, Aoyama Y (2007) Light-up Hoechst-DNA aptamer pair: generation of an aptamer-selective fluorophore from a conventional DNA-staining dye. ChemBioChem 8:1795–1803
27. Endo K, Nakamura Y (2010) A binary Cy3 aptamer probe composed of folded modules. Anal Biochem 400:103–109
28. Paige JS, Wu KY, Jaffrey SR (2011) RNA mimics of green fluorescent protein. Science 333:642–646
29. Kikuchi N, Kolpashchikov DM (2016) Split spinach aptamer for highly selective recognition of DNA and RNA at ambient temperatures. ChemBioChem 17:1589–1592
30. AlSalhi MS, Alam J, Dass LA, Raja M (2011) Recent advances in conjugated polymers for light emitting devices. Int J Mol Sci 12:2036–2054
31. Thomas SW, Joly GD, Swager TM (2007) Chemical sensors based on amplifying fluorescent conjugated polymers. Chem Rev 107:1339–1386
32. Gaylord BS, Heeger AJ, Bazan GC (2002) DNA detection using water-soluble conjugated polymers and peptide nucleic acid probes. Proc Natl Acad Sci USA 99:10954–10957
33. Pu F, Hu D, Ren JS, Wang S, Qu XG (2010) Universal platform for sensitive and label-free nuclease assay based on conjugated polymer and DNA/intercalating dye complex. Langmuir 26:4540–4545
34. Xu C, Zhou RY, Zhang RC, Yang LY, Wang GJ (2014) Label-free DNA sequence detection through FRET from a fluorescent polymer with pyrene excimer to SG. ACS Macro Lett 3:845–848
35. Chan WCW, Maxwell DJ, Gao XH, Bailey RE, Han MY, Nie SM (2002) Luminescent quantum dots for multiplexed biological detection and imaging. Curr Opin Biotechnol 13: 40–46
36. Eigen M, Rigler R (1994) Sorting single molecules-application to diagnostics and evolutionary biotechnology. Proc Natl Acad Sci USA 91:5740–5747
37. Lee J, Kim IS, Yu HW (2010) Flow cytometric detection of bacillus spoOA gene in biofilm using quantum dot labeling. Anal Chem 82:2836–2843
38. Yeh HC, Ho YP, Wang TH (2005) Quantum dot-mediated biosensing assays for specific nucleic acid detection. Nanomed-Nanotechnol Biol Med 1:115–121
39. Ho YP, Kung MC, Yang S, Wang TH (2005) Multiplexed hybridization detection with multicolor colocalization of quantum dot nanoprobes. Nano Lett 5:1693–1697
40. Zhang CY, Yeh HC, Kuroki MT, Wang TH (2005) Single-quantum-dot-based DNA nanosensor. Nat Mater 4:826–831
41. Algar WR, Krull UJ (2010) Multiplexed interfacial transduction of nucleic acid hybridization using a single color of immobilized quantum dot donor and two acceptors in fluorescence resonance energy transfer. Anal Chem 82:400–405
42. Algar WR, Krull UJ (2009) Interfacial transduction of nucleic acid hybridization using immobilized quantum dots as donors in fluorescence resonance energy transfer. Langmuir 25:633–638
43. Noor MO, Tavares AJ, Krull UJ (2013) On-chip multiplexed solid-phase nucleic acid hybridization assay using spatial profiles of immobilized quantum dots and fluorescence resonance energy transfer. Anal Chim Acta 788:148–157

44. Noor MO, Krull UJ (2014) Camera-based ratiometric fluorescence transduction of nucleic acid hybridization with reagentless signal amplification on a paper-based platform using immobilized quantum dots as donors. Anal Chem 86:10331–10339
45. Li M, Cushing SK, Wang QY, Shi XD, Hornak LA, Hong ZL, Wu NQ (2011) Size-dependent energy transfer between CdSe/ZnS quantum dots and gold nanoparticles. J Phys Chem Lett 2:2125–2129
46. Kang T, Kim HC, Joo SW, Lee SY, Ahn IS, Yoon KA, Lee K (2013) Optimization of energy transfer between quantum dots and gold nanoparticles in head-to-head configuration for detection of fusion gene. Sens Actuator B-Chem 188:729–734
47. Jenkins R, Burdette MK, Foulger SH (2016) Mini-review: fluorescence imaging in cancer cells using dye-doped nanoparticles. RSC Adv 6:65459–65474
48. Zhao XJ, Tapec-Dytioco R, Tan WH (2003) Ultrasensitive DNA detection using highly fluorescent bioconjugated nanoparticles. J Am Chem Soc 125:11474–11475
49. Montalti M, Prodi L, Rampazzo E, Zaccheroni N (2014) Dye-doped silica nanoparticles as luminescent organized systems for nanomedicine. Chem Soc Rev 43:4243–4268
50. Zhou XC, Zhou JZ (2004) Improving the signal sensitivity and photostability of DNA hybridizations on microarrays by using dye-doped core-shell silica nanoparticles. Anal Chem 76:5302–5312

# Chapter 8
# Electrochemical Sandwich Assays for Nucleic Acid Detection

**Meihua Lin and Xiaolei Zuo**

**Abstract** Quantitative determination of nucleic acids related to human health and safety has attracted a great interest. Electrochemical sandwich-type biosensor with simple operation and low price shows high sensitivity and specificity with dual recognition mechanism and has been widely used for nucleic acid detection. In this chapter, we highlight the advancements of electrochemical sandwich assay for nucleic acids in recent decade. We first introduced the importance of nucleic acid detection and the principles of design an electrochemical nucleic acid sandwich assay and then summarized the advancements of this strategy based on the types of reporter tags, including redox molecules, enzymes, and nanoparticles. Finally, we discussed the challenges in the development of electrochemical nucleic acid sandwich assay to apply for clinical diagnostics, in cells and in vivo.

**Keywords** Sandwich electrochemical biosensor · Nucleic acid detection
Label-free strategy · Redox label · Enzyme amplification · Nanoparticle
application · Engineering interface

---

The original version of this chapter was revised: Foreword has been included and authors' affiliations have been updated. The erratum to this chapter is available at https://doi.org/10.1007/978-981-10-7835-4_13

---

M. Lin
Engineering Research Center of Nano-Geomaterials of Ministry of Education,
Faculty of Materials Science and Chemistry, China University of Geosciences,
Wuhan 430074, People's Republic of China
e-mail: linmh@cug.edu.cn

X. Zuo (✉)
Institute of Molecular Medicine, Renji Hospital, School of Medicine and School
of Chemistry and Chemical Engineering, Shanghai Jiao Tong University,
Shanghai 200127, People's Republic of China
e-mail: zuoxiaolei@sjtu.edu.cn

## 8.1   Introduction

Nucleic acids (DNA, RNA) are biomolecules and play an important role in biological and physiological functions, including as genetic information carriers and transmitters, biochemical activities regulators, and reaction catalysts in cells and living organisms [1]. An abnormal expression of nucleic acids in cells, tissues, or serum is often associated with certain disease. For example, the most common form of genetic variation named single nucleotide polymorphism (SNP) with one point mutation in DNA, such as *KRAS* gene, is associated with lung cancer, colorectal cancer, and ovarian cancer [2]. And nucleic acids including RNA, microRNA (miRNA), and DNA, circulating in biofluids, such as blood, urine, and peritoneal fluid in patients, have been correlated with the severity and progression of breast cancer, prostate cancer, and bladder cancer [1]. Therefore, nucleic acids are a promising source of biomarkers for the early diagnosis of genetic diseases, effective monitoring drug efficacy, and prognosis after surgery treatment.

Detection of nucleic acids is critical in a variety of areas, including clinical diagnostics, gene therapy, environmental analysis, food safety monitoring, and antibiodefense [3]. Although there are many biological challenges for nucleic acid detection, especially in clinical samples and single cells, including low concentrations of nucleic acid biomarkers in clinical samples and complex matrix caused high interferences, a great deal of effort has been devoted to develop new biotechnologies in the last two decades, to detect nucleic acids in complex systems with high specificity and sensitivity. These methodologies mainly rely on highly specific Watson–Crick base pairing between nucleic acid strands, including polymerase chain reaction (PCR), quantitative reverse transcription PCR (qRT-PCR), fluorescence, colorimetry, Raman spectroscopy, Northern blotting, microarray, sequencing and electrochemical approaches [4]. PCR is the most commonly used method for amplification of DNA with ultrahigh sensitivity, achieving the detection limits to femtomolar or attomolar. However, the detection process is complicated, requiring rational design of primers, and thermal cycling equipment to achieve amplification, and is thus not suitable for rapid diagnostics [5]. qRT-PCR is the current gold standard for miRNA analysis by two steps. First, RNA is converted into complementary DNA (cDNA) by a reverse transcriptase. Then, the cDNA is amplified by DNA polymerase during PCR process, and the amplified cDNA can be quantified in real time by fluorescence or radioactivity signal. However, because of the short sequence of miRNA (approximately 19–23 nucleotides), the very short primers should be designed, which can easily produce a false-positive signal and affect the PCR efficiency due to a very low melting temperature [6]. It is also difficult to measure multiple miRNAs from within a single qRT-PCR reaction volume and needs internal controls, such as house-keeping genes to alleviate the variability of PCR amplification [7]. Due to the capability of short analytical time, cost effective, simple, sensitive, specific, high amenable to miniaturization and portable, and high potential to be multiplexed, electrochemical detection methods have attracted a lot of interest in the detection of nucleic acids.

In recent years, several review papers have been published by various groups on the development of electrochemical biosensor for biomolecules. Kelley group reviewed recent development of electrochemical methods for clinically relevant biomolecules [1], and Lin and colleagues discussed recent advances of electrochemical biosensor based on nanomaterials and nanostructures [8], whereas Hsing et al. summarized the combinations of the different strategies to get some generalized conclusions for designing a point-of-care electrochemical DNA biosensor [9]. However, there were few papers reviewed in the area of electrochemical sandwich assay [10]. Herein, in this chapter, we focus on the developments and challenges of electrochemical sandwich assay for nucleic acids, including DNA, RNA, and miRNA, and provide an update on the recent advances in this field to our last review in *Chemical Reviews*. For a clear illustration, the basic principles of electrochemical nucleic acid sandwich assay are firstly introduced and then the sensing strategies are classified based on the types of signal tags and divided into three categories: redox labels, enzyme amplification, and nanomaterials amplification.

## 8.2 Basic Principles of Electrochemical Sandwich Assays for Nucleic Acid Detection

Electrochemical sensing strategies employing nanostructured surfaces and redox signal amplifications have been widely used for nucleic acids in clinical diagnosis with high sensitivity and specificity. The sandwich assay is a basic method and plays a critical role in the detection of nucleic acids. Typically, an electrochemical sandwich-type assay contains three elements: capture probe, target probe, and signal probe, as shown in Fig. 8.1. First, the capture probe is immobilized on the electrode surface through chemical reaction or physical adsorption. With the development of DNA and RNA synthesis technology, it is easy to modify oligonucleotide sequence with many kinds of chemical molecule, such as thiol, carboxyl group, biotin, fluorescent dyes, redox molecules. Through these, the thiol modification capture probe can be self-assembled on the gold surface through Au–S bond, and carboxylic acid-labeled capture probe can react with hydroxyl group on

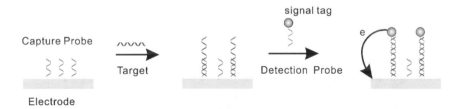

**Fig. 8.1** Schematic illustration of a typical electrochemical sandwich assay for nucleic acid analysis

the surface of glassy carbon electrode and indium tin oxide electrode. Then, target probe acts as a bridge, one part hybridization with capture probe, the other part hybridization with detection probe. Finally, the detection probe labeled with signal tags, such as methylene blue, ferrocene, and nanoparticles, produces electrochemical signals by many kinds of electrochemical technology, such as square-wave voltammetry, cyclic voltammetry, differential pulse voltammetry, anodic stripping voltammetry, chronocoulometry, and impedance spectroscopy.

Traditionally, the capture probe is single-strand DNA and decorated on bare electrode. Also, the detection probe is only tagged one molecule to produce electrochemical signal, which limits the sensitivity of the nucleic acid sandwich assay. In recent years, significant advances in DNA nanobiotechnology and nanomaterials provided new opportunities to create a new type of biosensor with much higher sensitivity and specificity. Because of the unique properties of nanomaterials, such as excellent conductivity, high surface area, advanced catalytic properties, finely tunable size and optical properties, and good biocompatibility, they have been introduced into the areas of electrochemical nucleic acid biosensor. However, in this chapter, we aim to summarize the electrochemical sandwich assay for nucleic acid detection based on the types of signal tag, including redox labels, enzymes, and nanoparticles, which show direct electrochemistry or catalytic properties. Therefore, nanomaterials which are conjugated with biomolecules and served as electrode materials and carriers to enhance capture probe immobilization and signal tag loaded for electrochemical signal amplification will be integrated into each section.

## 8.3  Application of Redox Labels

### 8.3.1  Label-Free Electrochemical Assays

Redox molecules are most used directly to produce an electrochemical signal for quantitative electrochemical measurements. Some of redox molecules can interact with nucleic acids without any covalent reaction; for example, hexaammineruthenium (III) chloride ($[Ru(NH_3)_6]^{3+}$, RuHex) can quantitatively bind to nucleic acid phosphate backbone by electrostatic interactions [11, 12] and methylene blue can bind strongly to nucleic acid sequence through intercalation [13]. They provide a label-free, simple, and cost-effective strategy for electrochemical detection of nucleic acids. Chunhai Fan and his co-worker developed a sandwich-type assay based on the redox charge of RuHex to detect DNA [11]. As shown in Fig. 8.2, capture probe was first immobilized on the gold electrode surface, and a short chain 6-mercapto-1-hexanol (MCH) was backfilled the surface to help capture probe "stand up," which can improve the recognition capability of capture probe. Then, the capture probe hybridized with target probe, along with the reporter DNA-AuNP conjugates. Because one gold nanoparticle carried several hundreds of reporter DNA which can adsorb thousands of RuHex, the detection limit was as low as

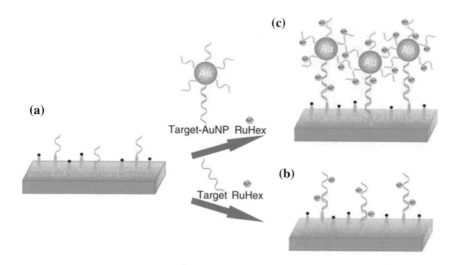

**Fig. 8.2** Schematic of a label-free sandwich strategy for DNA detection based on gold nanoparticle amplification. **a** Thiol-labeled capture probe and passivation molecule MCH are immobilized on a gold electrode. **b** Target probe hybridizes with capture probe without gold nanoparticle amplification. **c** Target probe along with reporter probe–gold nanoparticle conjugates hybridizes with capture probe, producing an amplification signal by adsorbing thousands of $Ru^{3+}$ (Reprinted with the permission from Ref. [11]. Copyright 2007 Nature Publishing Group)

femtomolar. Zhang et al. further improved the detection sensitivity by 9.2 times through fabrication of nanoporous gold electrode [12]. They achieved 28 aM sensitivity and the linear range of 80 aM–1.6 pM with excellent selectivity for single-mismatched DNA detection. This strategy was also applied for human telomerase analysis [14]. Telomerase extracted from HeLa cells extended the telomerase substrate primer, which acted as a sandwich-type target probe, bringing the capture probe and signal probe. They demonstrated that this gold nanoparticle amplification method is sensitive, no need of PCR, and could detect the telomerase activity from about 10 cultured cancer cells. Because each $Ru^{3+}$ can only accept one electron, the sensitivity is insufficient to detect clinical samples. Kelley and her co-workers created a significant signal amplification method by introducing a second electron acceptor $Fe(CN)_6^{3-}$, which acts as an oxidant to regenerate $Ru^{3+}$ [15]. They found the current was increased by 10–50 times, and this amplification assay was successfully used for direct analysis of mutated sequences rapidly in patient serum [16, 17]. Wang and his colleagues have further improved the detection sensitivity and stability of miRNA-21 by preparation gold nanoparticle-decorated $MoS_2$ nanosheet as a work electrode surface and reporter probe carrier to increase the amount of probes [18].

Although RuHex is a good electrochemical reporter, RuHex is expensive and nonspecific to single-stranded (ss) and double-stranded (ds) DNA. Much cost-effective redox markers were explored, for example, methylene blue,

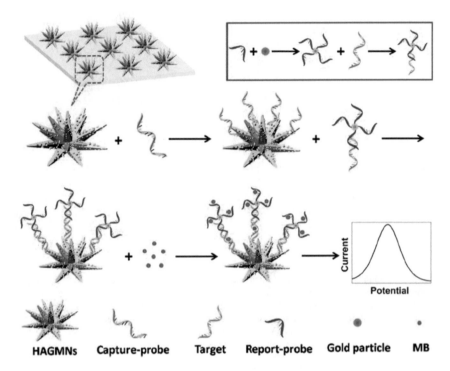

**Fig. 8.3** Schema of DNA sandwich assay performed on a hydrophilic three-dimensional aloe-like gold micro/nanostructures electrode with increase amount of capture probe. Also, the signal was amplified by gold nanoparticles loading many reporter probes, which could be inserted by more amount of methylene blue molecules (Reprinted with the permission from Ref. [21]. Copyright 2013 Elsevier)

adriamycin [19]. Methylene blue is a phenothiazine dye and performs different affinity to ss- and dsDNA nucleic acids as an electrochemical indicator [20]. Yao et al. reported an electrochemical biosensor for ultrasensitive miRNA detection by using methylene blue as indicator [13]. Under optimal conditions, the detection limit of 0.5 fM was achieved. To improve the detection sensitivity of methylene blue-based sandwich, Shi and co-workers fabricated hydrophilic three-dimensional aloe-like gold micro/nanostructures (HAG) with large effective area for immobilizing more capture probe and combined with gold nanoparticle amplification for DNA detection (Fig. 8.3) [21]. Gold nanoparticles carried about one hundred reporter probes, which would adsorb a large number of methylene blue molecules, leading to a significant increased current. On the basis of this amplification, they achieved a detection limit of 12 aM, with a linear range from 50 aM to 1 pM.

## 8.3.2   Redox-Labeled Electrochemical Assays

The development of the technology of DNA and RNA synthesis promotes the capability of oligonucleotide conjugated to fluorophore, enzyme, nanoparticle, and redox molecule. Redox-labeled electrochemical sandwich assay has been produced, such as labeled with methylene blue [22], ferrocene [23, 24], anthraquinone [25]. Compared to label-free intercalator methylene blue, methylene blue conjugated to oligonucleotide is more sensitive to SNP and provides more efficient electron transfer with good stability and can be used in serum and whole blood [26–28]. Li et al. developed a sandwich assay for *Escherichia coli* (*E. coli*) *lac Z* gene detection with DNA-attached methylene blue as the detection probe [29]. This method successfully detected *E. coli* gene sequence with sensitivity of about 30 fM and specific discrimination mismatch DNA. Moradi et al. also created a sandwich assay for discrimination of single-base mismatch DNA based on oxidization covalently conjugated ferrocenecarboxylic acid [30]. Because each redox has a unique potential, it has a potential to develop a multiplexed detection assay and ratiometric assay. Cheeveewattanagul et al. constructed a sandwich assay for the simultaneous detection of four different targets [31]. As shown in Fig. 8.4, the four different reporter probes were tagged silica nanoparticles, which were loaded for four different redox molecules, methylene blue, ferrocenium tetrafluoroborate, tris (2,2-bipyridyl) dichlororuthenium (II) hexahydrate, and acridine orange via electrostatic interaction, producing four different potentials. They achieved picomolar sensitivity for simultaneous detection of four different sequences in four types of the influenza virus.

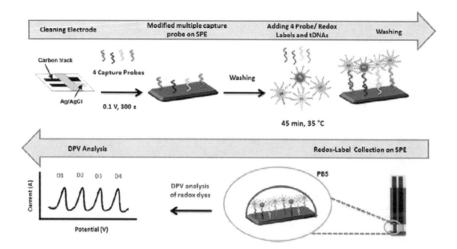

**Fig. 8.4** Schema of the simultaneous DNA sandwich assay, by immobilization of four different capture probes onto screen-printed electrodes, and amplification by silica nanoparticles loaded with four different redox molecules, including methylene blue, ferrocenium tetrafluoroborate, tris (2,2-bipyridyl) dichlororuthenium (II) hexahydrate, and acridine orange (Reprinted with the permission from Ref. [31]. Copyright 2016 Elsevier)

## 8.4 Application of Enzyme-Based Amplification

### 8.4.1 Horseradish Peroxidase Amplification Assays

Each redox molecule mentioned above often only transfers one or a few electrons, limiting the sensitivity of nucleic acid detection. Many enzyme-linked DNA sandwich assays due to their fast response, ultrahigh sensitivity, and selectivity have been developed to amplify the electrochemical signals for the detection of pathogen DNA, cancer relative DNA, mRNA, and miRNA. During this amplification, each enzyme can catalyze thousands of reactions, causing thousands of electrons transferring to the electrode surface. Horseradish peroxidase (HRP) is most extensively used to generate significant signals in the sandwich assay. HRP is kind of enzyme that can catalyze hydrolyzation of $H_2O_2$, producing electrons which cannot directly transfer to the electrode. There are many methods employed to wire the electrons to the electrode surface. 3,3',5,5'-tetramethylbebzidine (TMB) is most widely used as an electron shuttle of the reduction of hydrogen peroxide by HRP to the electrode surface [32–34]. Since it is easy to synthesize oligonucleotides labeled with biotin, avidin- or streptavidin-HRP is mostly introduced to act as a signal tag. On the basis of these, Fan's group has constructed a few sandwich assays for pathogen DNA, cancer cell miRNA detection [32–37]. As shown in Fig. 8.5, by coassembled with protein-resistant oligo(ethylene glycol) (OEG)-terminated thiols, capture probe performed nonfouling feature and detected target DNA in human

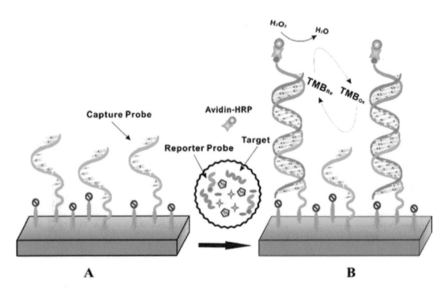

**Fig. 8.5** Schema of DNA sandwich assay based on HRP. Here, OEG was incorporated with capture probe as backfiller and protein resistant (Reprinted with the permission from Ref. [35]. Copyright 2008 American Chemical Society)

serum with a detection limit of 1 pM. This OEG passivation layer does not only significantly eliminate nonspecific DNA adsorption and stand up the capture probes as MCH, but also possess high repel to nonspecific adsorption of proteins [35]. Then, this system was extended to discriminate target DNA from different SNPs simultaneously by the use of a 16-electrode sensor array [36].

To construct more compact surface monolayer, Joseph Wang and his co-workers created a new ternary surface monolayer (including thiolated capture DNA probe, MCH, and dithiothreitol) which reduced surface detects, minimized the background contributions, and then increased the signal-to-noise ratio [38]. The detection limit was as low as 40 zmol (in 4 μl samples) as well as 1 CFU *E. coli*, by combining this new ternary monolayer with HRP amplification. Although this method is supersensitive, the assembled process is complex, multistep, time-consuming, and reagent-consuming. Chen and his colleagues designed a bovine serum albumin (BSA)-monolayer-based platform to better control the interspace between immobilization capture probes to enhance the accessibility of the target probes (Fig. 8.6) [39]. Because the high resistant ability to nonspecific adsorption of protein, the BSA-based biosensor can be directly applied in biological fluids. This fast, simple immobilization approach based on a sandwich assay by using HRP as a signal tag performed high reproducibility and obtained a detection limit of 0.5 fM, which was 500,000-fold improvement than MCH-based conventional biosensor.

However, BSA-based probe carrier platform can increase the interspace between neighboring capture probes and the flexibility of ssDNA will cause capture probe nonspecific interaction with surface, restricting the accessibility of target probe. In addition, the BSA-based platform increases the heterogeneity of the biosensor surface. DNA nanotechnology provides new opportunities to engineer-ordered and

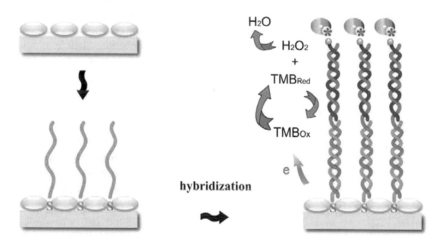

**Fig. 8.6** BSA-based DNA sandwich assay is amplified by HRP with high reproducibility and sensitivity (Reprinted with the permission from Ref. [39]. Copyright 2013 American Chemical Society)

well-controllable biomolecule-confined biosensing interfaces to increase the ability of molecule recognition. As an example, Chunhai Fan's group has firstly introduced well-defined three-dimensional tetrahedral DNA nanostructures (TDNs) into electrode surface to develop various biosensors, including DNA, miRNA, protein, small molecule, and cell sensors [40]. Typically, a tetrahedral DNA nanostructure is self-assembled by equimolar quantities of four single-stranded DNA sequences by heating at 95 °C for two minutes and fast cooling at 4 °C in 30 s for more than five minutes, with a high yield of over 85% [41]. A TDN with a pendant DNA probe at one vertex is anchored onto gold electrode surface by three thiol groups modified at the other three vertices. The process of self-assembled of the TDN on the gold surface was monitored and verified by quartz crystal microbalance, surface plasmon resonance, and microcantilever [33]. Since TDNs are highly rigid, the capture probe at the top vertex can keep an upright orientation with well-controlled spacing, avoiding interprobe entanglement, and increasing the molecular accessibility. Moreover, the uniform and thick layer of TDN provides a solution-phase-like environment for capture probe, improving the kinetics and efficiency of biomolecular recognition. Pei et al. first constructed a TDN-based DNA sandwich assay with HRP-based signal-transduction approach [33]. The sensitivity for DNA detection was 1 pM without the requirement of "help" molecules, such as MCH, OEG, exhibiting a 250-fold improvement compared with traditional ssDNA probe. The TDN modification surfaces have an ability to prevent nonspecific adsorption of protein and can directly perform in biological fluids with excellent discrimination ability for SNP. This TDN-based electrochemical platform also provides an opportunity for sensitive detection of miRNAs. Wen et al. designed a sandwich assay for sensitive and specific detection of miRNA 21, which is overexpressed in many kinds of cancer cells (Fig. 8.7) [37]. A short capture DNA probe appended at the top vertex of the tetrahedron structure, and a biotin-labeled signal probe, flanked the target miRNA. The detection limit can be lower than 10 fM by employing avidin-HRP as a catalyst, while the sensitivity will be improved 100-fold, as low as 10 aM, by employing poly-HRP80 (a polymerized streptavidin-HRP conjugate with up to 400 HRP molecules) as a catalyst. And this protein-resistant TDN surface not only performed high signal-to-background ratio, but also exhibited excellent recognition ability of target miRNA in real, complex clinical samples, being a promising miRNA analysis method in clinical diagnosis. Multiplex miRNAs detection was achieved by employing this TDN-based sandwich assay with 16-channel screen-printed gold electrode [42]. Four pancreatic carcinoma-related miRNAs were detected simultaneously with detection limits of 10 fM. To improve biosensing performance, Fan's group continued developing a programmable "soft lithography" approach for interface engineering by five different sizes of TDNs [34]. The biosensor performance is regulated by changing the size of TDN. By increasing the TDN size, the hybridization rate was 20-fold improvement, and the hybridization efficiency was fivefold enhancement, compared with ssDNA probe. And the detection limit was also tunable by integrating DNA

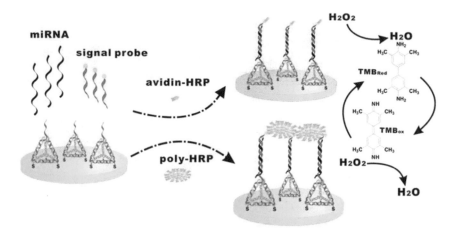

**Fig. 8.7** Schematic illustration of TDN-based sandwich assay for miRNA analysis. The biotin-tagged reporter probe can specifically bind to avidin-HRP or polymerized streptavidin-HRP conjugates to produce catalytic signal (Reprinted with the permission from Ref. [37]. Copyright 2012 Nature Publishing Group)

sandwich assay. With the increasement of lateral distance, the detection limits were lowed from 10 pM to 1 fM, and the detection limit of larger size could be more sensitive and get a limit to 100 aM by using poly-HRP40 as a catalyst.

With the achievements in nanotechnology and nanoscience, inorganic nano-materials acting as electrode materials or enzyme carriers to produce a synergic effect for signal amplification have been extensively used in electrochemical sandwich assays. He and his colleagues employed a $SiO_2$ nanoparticle as a carrier wrapped with streptavidin-HRP for sensitive DNA detection by a sandwich assay [43]. Liu et al. modified the glassy carbon electrode with graphene–three-dimensional nanostructure gold nanocomposite for increasing the active surface area to immobilize more capture probes [44]. The detection limit for target DNA was as low as 3.4 fM, with a linear range of 50–5000 fM, by employing a HRP-based sandwich strategy. If the electrode was firstly modified with nanomaterial, and then combined with nanomaterial loaded with lots of enzymes, the detection sensitivity will be much more improved. Ju and his co-workers constructed an ultrasensitive sandwich assay by multiple modification and amplification [45]. As shown in Fig. 8.8, graphene oxide was firstly modified on a glassy carbon electrode by electrochemical reduction and then thiol-labeled ssDNA was assembled on the graphene oxide by the π–π stacking interaction. Then, capture probe-capped gold nanoparticles were anchored on the electrode surface by thiol-Au bond, and after biotin-tagged target DNA hybridized with capture probe, streptavidin-HRP-functionalized carbon nanosphere (CNS) bound specifically to the target. This multiplex signal amplification strategy achieved an attomolar-level detection limit with a linear range from 100 fM to 10 aM and exhibited a high selectivity to SNPs. Zhou et al. developed a similar strategy by employing

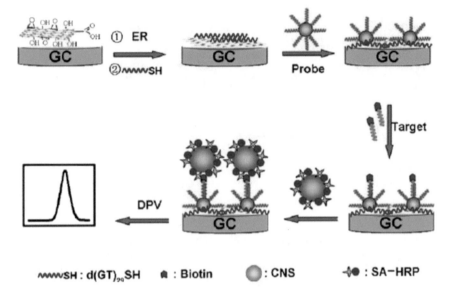

**Fig. 8.8** Schema of representative multiplex amplification of electrochemical sandwich strategy by employing nanomaterials as carriers of both capture probes and signal enzymes (Reprinted with the permission from Ref. [45]. Copyright 2012 Elsevier)

mesoporous carbon nitride and gold nanoparticle modification electrode and gold nanocluster-loaded streptavidin-HRP [46]. The detection limit was also very low, achieved 8 aM. Although these nanomaterial modification sandwich assays are very sensitive, the fabrication process is always multistep and complicated.

## 8.4.2   Other Enzyme Amplification Assays

Besides the contribution of HRP in the enzyme-based sandwich strategy, alkaline phosphatase (ALP) and glucose oxidase also have been widely explored in the electrochemical sandwich assay. ALP could selectively mediate precipitation of an insoluble and insulating product and has been used in a scanning electrochemical microscopy feedback signal for target DNA detection [47]. While this detection sensitivity was very poor, Joseph Wang and his colleagues created an electro-chemical sandwich based on ALP-induced metallization [48]. ALP catalyzed p-Aminophenyl phosphate monosodium salt hydrate (p-APP) to p-aminophenol (p-AP), which reduced silver ions to Ag. The potential change of silver ion-selective electrode ($Ag^+$-ISE) caused by the decrease of silver ions could be used as the hybridization signal. The detection limits of 50 fM target DNA and 10 CFU in the 4 μL sample were obtained, respectively. Lin and his co-workers further improved the detection sensitivity by using branched DNA loading high amount of ALP tracers [49]. This sandwich assay for PCR-free detection RNA reached a detection limit of 1 fM. Nanomaterials also have been used in ALP-based electrochemical sandwich assays [50, 51]. For example, Shuai et al. constructed a sandwich strategy by using magnesium oxide nanoflower and gold nanoparticles as electrode surface materials, and graphene oxide-gold nanoparticles as scaffold for carrying many streptavidin-ALP-labeled reporter probes [50]. This biosensor performed well for miRNA detection with a detection limit of 50 aM and broad dynamic range from 0.1 to 100 fM.

Glucose oxidase-based sandwich assays have successfully used for the detection of DNA and miRNA [52–54]. As an example, Fan and his colleagues built a nonfouling electrode surface by self-assembled thiolated capture probe and OEG to capture a specific DNA sequence [53]. Glucose oxidase tagged on reporter probe catalyzed glucose to produce hydrogen peroxide and generate electrochemiluminescent signal. This sandwich-type DNA sensor has a good performance in complicated biological fluids and shows a detection limit of 1 pM.

## 8.5  Application of Nanoparticles

### 8.5.1  Electrocatalytic Assays

Although enzyme-based sandwich assays are ultrasensitive and have been widely used, they have intrinsic limitations, such as poor thermal and environmental stability, expensiveness, and difficulty to transport and storage. Nanomaterials with unique optical, electronic, and catalytic properties provide an opportunity to construct a stable and sensitive electrochemical sandwich assay. Recently, many kinds of nanomaterials mimicking peroxidase have been attracted great interest as they have many advantages relative to natural enzymes, such as simple preparation, cost-effectiveness, good stability, and inertness to several proteases. Zhang and his colleagues prepared artificial peroxidase nanoparticles (ZrHCF MNPs), which were composed with magnetic beads inner core and zirconium hexacyanoferrate(II) outer shell, and incorporated with reporter probe through zirconium-(OPO3-poly(dG) DNA) covalent bonds (Fig. 8.9) [55]. By employing a simple sandwich strategy, target DNA as low as 0.43 fM with a linear range from 1 fM to 1 nM has been detected by the particles catalyzing $H_2O_2$. Miao et al. also constructed a sandwich strategy for miRNA detection by using an iridium(III) complex as a peroxidase-like mimic [56].

Besides, noble metals exhibiting mimic enzyme properties have been realized. Silver nanoclusters, gold nanoparticles, and Pt nanoparticles displayed efficient peroxidase properties for the reduction of hydrogen peroxide and have employed in sandwich assays for detection of target miRNA and DNA [57–59]. To further improve the electrochemical catalytic current of Pt nanoparticles, Xu and her colleagues fabricated DNAzyme-functionalized Pt nanoparticle/carbon nanotube

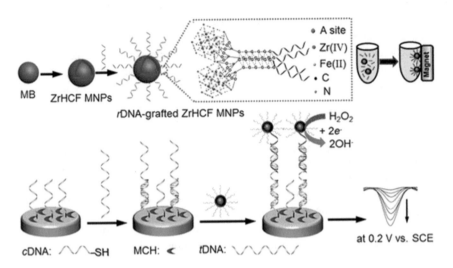

**Fig. 8.9** Schema of the preparations of reporter DNA incorporated with ZrHCF MNPs and schematic illustration of the DNA sandwich assay based on ZrHCF MNPs catalyzing the reduction of $H_2O_2$ (Reprinted with the permission from Ref. [55]. Copyright 2015 American Chemical Society)

bioconjugate to label the reporter DNA through platinum–sulfur bond and constructed a sandwich assay for DNA detection with a detection limit of 0.6 fM [60]. Zhang and his co-workers also improved the electrochemical catalytic ability of Pt nanoparticles by decorating very small size of Pt nanoparticles on the surface of tin-doped indium oxide nanoparticles and forming Pt/Sn-In$_2$O$_3$ [61]. The superior catalytic behaviors of Pt nanoparticles are not only the catalysts for the reduction of hydrogen peroxide and oxygen, but also the catalyst for the oxidation of hydrazine. Bard et al. demonstrated a proof-of-concept study on the behavior of single Pt nanoparticle for low concentration of DNA detection by a sandwich strategy (Fig. 8.10) [62]. In the present high concentration of target DNA, it was hard to distinguish the electrocatalytic current from the background, while in the present low concentration of target DNA (10 pM), an electrocatalytic response was monitored by single Pt nanoparticle collisions with an ultramicroelectrode (UME) for the electrochemical oxidation of hydrazine. The ability of detection individual biomolecules offers future a promising in the development of more sensitive sensors.

## 8.5.2 Direct Electroactive Assays

Nanoparticles also can perform directly electrochemistry properties, such as redox active, and be used as signal tracers in electrochemical sandwich assays. For

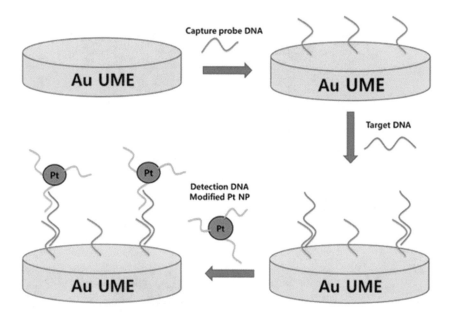

**Fig. 8.10** Schema of the DNA sandwich assay based on single Pt nanoparticles collisions at an Au UME surface. At low concentration of DNA, individual Pt nanoparticles collision events were monitored by employing Pt nanoparticle/Au UME/hydrazine oxidation reaction (Reprinted with the permission from Ref. [62]. Copyright 2012 American Chemical Society)

example, gold nanoparticle labeled on the reporter probe in a sandwich assay can dissolve as $Au^{3+}$ by stripping voltammetry and produce electrochemical signal [63]. A novel DNA sensor based on gold nanoparticles-catalyzed silver deposition has been developed [64, 65]. Li and his colleagues fabricated an electrochemical sandwich-type DNA sensor by assembling capture probe on the graphene-modified glassy carbon electrode surface and introducing reporter probe on the gold nanoparticle surface [65]. After target DNA hybridized with capture probe and reporter probe, silver deposition on the gold nanoparticle surface was carried out. The deposited silver was then detected by electrochemical stripping technique to generate an amplified current signal. As a result, the detection limit of 72 pM was obtained. To enhance the signal amplification, Chen and his co-workers made an aggregated Ag nanostructure as a reporter probe tag (Fig. 8.11) [66]. As shown in Fig. 8.11a, reporter probe and polyA were modified on the silver nanoparticles firstly. Then, these nanoparticles hybridized with the silver nanoparticles modified with polyT form aggregated nanostructures, 10 times larger than the individual nanoparticle. Based on a sandwich assay, the sensitivity of silver nanoparticle aggregate label was 1000-fold higher than a single nanoparticle label, achieving 5 aM. Moreover, this strategy was successful to detect multiplexed target DNA by using array chips (Fig. 8.11b).

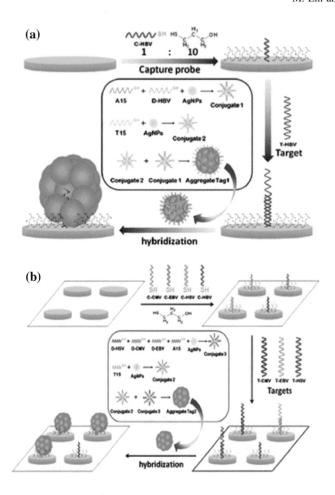

**Fig. 8.11** **a** Schema of the DNA sandwich based on Ag nanoparticle aggregates. **b** Schema of multiplex detection based on array chips (Reprinted with the permission from Ref. [66]. Copyright 2010 American Chemical Society)

Quantum dots with size-tunable properties and high mechanical and thermal stability often contain metal oxide or metal sulfide and can be used as electroactive labels. A layer-by-layer self-assembly of CdTe quantum dots on polybeads has been tagged with DNA probes for sensitive DNA detection [67]. A detection limit of 0.52 fM was obtained by the detection of $Cd^{2+}$ dissolved from CdTe quantum dots, based on a sandwich assay. Since quantum dots have different stripping peaks, they provide multiplexed capability for sensitively simultaneous target analysis. Three quantum dots (PbS, CdS, and ZnS) were labeled with three different DNA sequences as reporter probes [68]. In a sandwich assay, these reporter probes hybridized with their target sequence and generated three distinct peaks by stripping voltammetry. This multiplex detection is a promising strategy for clinical diagnostics for simultaneous detection of several biomarkers.

Another ultrasensitive sandwich-type biosensor was developed based on electrochemical current response of $Cd^{2+}$ [69]. Reporter probe was modified with biotin, which can specifically bind to streptavidin–titanium phosphate nanoparticle–$Cd^{2+}$. In the presence of the target miRNA, a sandwich complex was formed. Upon addition of $Ru(NH3)_6^{3+}$ that bound to nucleic acid phosphate residues through electrostatic interaction and acted as electron wire, the electrochemical current increased significantly. The biosensor showed a very low detection limit of 0.76 aM with a linear range from 1 aM to 10 pM and has ability to discriminate the target miRNA from single-mismatched miRNA sequences. What's more, this strategy performed well in response to target miRNA in the human serum, which provides an opportunity to detect miRNA in clinical samples.

## 8.6 Conclusion

In this chapter, we have summarized the recent development of electrochemical sandwich assay for nucleic acid analysis. Benefitted from the fast development of DNA nanotechnology, DNA synthesis technology, chemical modification, and nanomaterial synthetic methodologies, the electrochemical interface and the signal reporters have been significantly improved to promote the target hybridization, electron conduction, and signal amplification. The achievement of high sensitivity, selectivity, and protein resistant of many nucleic acid sandwich strategies offers a promising future for clinical diagnostics. Moreover, by the development of microfabrication and microelectronic technique, it is possible to combine the microfluidic with electrochemistry to construct a miniaturized biosensor device for the point-of-care detection in real samples. Also, we still need to make efforts to design a reagent-free and wash-free electrochemical sandwich assay to apply in real-time detection in cells and in vivo.

## References

1. Labib M, Sargent EH, Kelley SO (2016) Electrochemical methods for the analysis of clinically relevant biomolecules. Chem Rev 116:9001–9090
2. Huber F, Lang HP, Backmann N, Rimoldi D, Gerber C (2013) Direct detection of a BRAF mutation in total RNA from melanoma cells using cantilever arrays. Nat Nanotechnol 8:125–129
3. Song SP, Qin Y, He Y, Huang Q, Fan CH, Chen HY (2010) Functional nanoprobes for ultrasensitive detection of biomolecules. Chem Soc Rev 39:4234–4243
4. Du Y, Dong SJ (2017) Nucleic acid biosensors: recent advances and perspectives. Anal Chem 89:189–215

5. Zhou W, Gao X, Liu DB, Chen XY (2015) Gold nanoparticles for in vitro diagnostics. Chem Rev 115:10575–10636
6. Harvey JD, Jena PV, Baker HA, Zerze GH, Williams RM, Galassi TV, Roxbury D, Mittal J, Heller DA (2017) A carbon nanotube reporter of microRNA hybridization events in vivo. Nat Biomed Eng 1:0041
7. Graybill RM, Bailey RC (2016) Emerging biosensing approaches for microRNA analysis. Anal Chem 88:431–450
8. Zhu CZ, Yang GH, Li H, Du D, Lin YH (2015) Electrochemical sensors and biosensors based on nanomaterials and nanostructures. Anal Chem 87:230–249
9. Yu HLL, Maslova A, Hsing IM (2017) Rational design of electrochemical DNA biosensors for point-of-care applications. ChemElectroChem 4:795–805
10. Shen JW, Li YB, Gu HS, Xia F, Zuo XL (2014) Recent development of sandwich assay based on the nanobiotechnologies for proteins, nucleic acids, small molecules, and ions. Chem Rev 114:7631–7677
11. Zhang J, Song SP, Wang LH, Pan D, Fan CH (2007) A gold nanoparticle-based chronocoulometric DNA sensor for amplified detection of DNA. Nat Protoc 2:2888–2895
12. Hu KC, Lan DX, Li XM, Zhang SS (2008) Electrochemical DNA biosensor based on nanoporous gold electrode and multifunctional encoded DNA–Au bio bar codes. Anal Chem 80:9124–9130
13. Li FY, Peng J, Zheng Q, Guo X, Tang H, Yao SZ (2015) Carbon nanotube-polyamidoamine dendrimer hybrid-modified electrodes for highly sensitive electrochemical detection of microRNA24. Anal Chem 87:4806–4813
14. Li Y, Liu BW, Li X, Wei QL (2010) Highly sensitive electrochemical detection of human telomerase activity based on bio-barcode method. Biosens Bioelectron 25:2543–2547
15. Sage AT, Besant JD, Lam B, Sargent EH, Kelley SO (2014) Ultrasensitive electrochemical biomolecular detection using nanostructured microelectrodes. Acc Chem Res 47:2417–2425
16. Das J, Ivanov I, Montermini L, Rak J, Sargent EH, Kelley SO (2015) An electrochemical clamp assay for direct, rapid analysis of circulating nucleic acids in serum. Nat Chem 7:569–575
17. Das J, Ivanov I, Sargent EH, Kelley SO (2016) DNA clutch probes for circulating tumor DNA analysis. J Am Chem Soc 138:11009–11016
18. Su S, Cao WF, Liu W, Lu ZW, Zhu D, Weng LX, Wang LH, Fan CH, Wang LH (2017) Dual-mode electrochemical analysis of microRNA-21 using gold nanoparticle-decorated $MoS_2$ nanosheet. Biosens Bioelectron 94:552–559
19. Shi AQ, Wang J, Han XW, Fang X, Zhang YZ (2014) A sensitive electrochemical DNA biosensor based on gold nanomaterial and graphene amplified signal. Sens Actuators B 200:206–212
20. Boon EM, Ceres DM, Drummond TG, Hill MG, Barton JK (2000) Mutation detection by electrocatalysis at DNA-modified electrodes. Nat Biotechnol 18:1096–1100
21. Shi L, Chu ZY, Liu Y, Jin WQ, Chen XJ (2013) Facile synthesis of hierarchically aloe-like gold micro/nanostructures for ultrasensitive DNA recognition. Biosens Bioelectron 49:184–191
22. Xia F, White RJ, Zuo XL, Patterson A, Xiao Y, Kang D, Gong X, Plaxco KW, Heeger AJ (2010) An electrochemical supersandwich assay for sensitive and selective DNA detection in complex matrices. J Am Chem Soc 132:14346–14348
23. Liu SP, Su WQ, Li ZL, Ding XT (2015) Electrochemical detection of lung cancer specific microRNAs using 3D DNA origami nanostructures. Biosens Bioelectron 71:57–61
24. Ahangar LE, Mehrgardi MA (2012) Nanoporous gold electrode as a platform for the construction of an electrochemical DNA hybridization biosensor. Biosens Bioelectron 38:252–257
25. Chao J, Cao WF, Su S, Weng LX, Song SP, Fan CH, Wang LH (2016) Nanostructure-based surface-enhanced Raman scattering biosensors for nucleic acids and proteins. J Mater Chem B 4:1757–1769

26. Li H, Arroyo Currás N, Kang D, Ricci F, Plaxco KW (2016) Dual-reporter drift correction to enhance the performance of electrochemical aptamer-based sensors in whole blood. J Am Chem Soc 138:15809–15812
27. Kang D, Ricci F, White RJ, Plaxco KW (2016) Survey of redox-active moieties for application in multiplexed electrochemical biosensors. Anal Chem 88:10452–10458
28. Pheeney CG, Barton JK (2012) DNA electrochemistry with tethered methylene blue. Langmuir 28:7063–7070
29. Li FQ, Yu ZQ, Qu HC, Zhang GL, Yan H, Liu X, He XJ (2015) A highly sensitive and specific electrochemical sensing method for robust detection of *Escherichia coli* lac Z gene sequence. Biosens Bioelectron 68:78–82
30. Moradi N, Noori A, Mehrgardi MA, Mousavi MF (2016) Scanning electrochemical microscopy for electrochemical detection of single-base mismatches by tagging ferrocenecarboxylic acid as a redox probe to DNA. Electroanalysis 28:823–832
31. Cheeveewattanagul N, Rijiravanich P, Surareungchai W, Somasundrum M (2016) Loading of silicon nanoparticle labels with redox mediators for detection of multiple DNA targets within a single voltammetric sweep. J Electrochem Soc 779:61–66
32. Liu G, Wan Y, Gau V, Zhang J, Wang LH, Song SP, Fan CH (2008) An enzyme-based E-DNA sensor for sequence-specific detection of femtomolar DNA targets. J Am Chem Soc 130:6820–6825
33. Pei H, Lu N, Wen YL, Song SP, Liu Y, Yan H, Fan CH (2010) A DNA nanostructure-based biomolecular probe carrier platform for electrochemical biosensing. Adv Mater 22:4754–4758
34. Lin MH, Wang JJ, Zhou GB, Wu N, Lu JX, Gao JM, Chen XQ, Shi JJ, Zuo XL, Fan CH (2015) Programmable engineering of a biosensing interface with tetrahedral DNA nanostructures for ultrasensitive DNA detection. Angew Chem Int Ed 54:2151–2155
35. Zhang J, Lao RJ, Song SP, Yan ZY, Fan CH (2008) Design of an oligonucleotide-incorporated nonfouling surface and its application in electrochemical DNA sensors for highly sensitive and sequence-specific detection of target DNA. Anal Chem 80:9029–9033
36. Wan Y, Zhang J, Liu G, Pan D, Wang LH, Song SP, Fan CH (2009) Ligase-based multiple DNA analysis by using an electrochemical sensor array. Biosens Bioelectron 24:1209–1212
37. Wen YL, Pei H, Shen Y, Xi JJ, Lin MH, Lu N, Shen XZ, Li J, Fan CH (2012) DNA Nanostructure-based Interfacial engineering for PCR-free ultrasensitive electrochemical analysis of microRNA. Sci Rep 2:867–872
38. Wu J, Campuzano S, Halford C, Haake DA, Wang J (2010) Ternary surface monolayers for ultrasensitive (zeptomole) amperometric detection of nucleic acid hybridization without signal amplification. Anal Chem 82:8830–8837
39. Liu YH, Li HN, Chen W, Liu AL, Lin XH, Chen YZ (2013) Bovine serum albumin-based probe carrier platform for electrochemical DNA biosensing. Anal Chem 85:273–277
40. Lin MH, Song P, Zhou GB, Zuo XL, Aldalbahi A, Lou XD, Shi JY, Fan CH (2016) Electrochemical detection of nucleic acids, proteins, small molecules and cells using a DNA-nanostructure-based universal biosensing platform. Nat Protoc 11:1244–1263
41. Goodman RP, Schaap IAT, Tardin CF, Erben CM, Berry RM, Schmidt CF, Turberfield AJ (2005) Rapid chiral assembly of rigid DNA building blocks for molecular nanofabrication. Science 310:1661–1665
42. Zeng DD, Wang ZH, Meng ZQ, Wang P, San LL, Wang W, Aldalbahi A, Li L, Shen JW, Mi XQ (2017) DNA tetrahedral nanostructure-based electrochemical miRNA biosensor for simultaneous detection of multiple miRNAs in pancreatic carcinoma. ACS Appl Mater Interfaces 9:24118–24125
43. Fan HJ, Wang XL, Jiao F, Zhang F, Wang QJ, He PG, Fang YZ (2013) Scanning electrochemical microscopy of DNA hybridization on DNA microarrays enhanced by HRP-modified SiO$_2$ nanoparticles. Anal Chem 85:6511–6517
44. Liu AL, Zhong GX, Chen JY, Weng SH, Huang HA, Chen W, Lin LQ, Lei Y, Fu FH, Sun ZL, Lin XH, Lin JH, Yang SY (2013) A sandwich-type DNA biosensor based on

electrochemical co-reduction synthesis of graphene-three dimensional nanostructure gold nanocomposite films. Anal Chim Acta 767:50–58

45. Dong HF, Zhu Z, Ju HX, Yan F (2012) Triplex signal amplification for electrochemical DNA biosensing by coupling probe-gold nanoparticles–graphene modified electrode with enzyme functionalized carbon sphere as tracer. Biosens Bioelectron 33:228–232

46. Zhou YY, Tang L, Zeng GM, Chen J, Wang JJ, Fan CZ, Yang GD, Zhang Y, Xie X (2015) Amplified and selective detection of manganese peroxidase genes based on enzyme-scaffolded-gold nanoclusters and mesoporous carbon nitride. Biosens Bioelectron 65:382–389

47. Palchetti I, Laschi S, Marrazza G, Mascini M (2007) Electrochemical imaging of localized sandwich DNA hybridization using scanning electrochemical microscopy. Anal Chem 79:7206–7213

48. Wu J, Chumbimuni-Torres KY, Galik M, Thammakhet C, Haake DA, Wang J (2009) Potentiometric detection of DNA hybridization using enzyme-induced metallization and a silver ion selective electrode. Anal Chem 81:10007–10012

49. Lee AC, Dai ZY, Chen BW, Wu H, Wang J, Zhang AG, Zhang LR, Lim T, Lin YH (2008) Electrochemical branched-DNA assay for polymerase chain reaction-free detection and quantification of oncogenes in messenger RNA. Anal Chem 80:9402–9410

50. Shuai HL, Huang KJ, Zhang WJ, Cao XY, Jia MP (2017) Sandwich-type microRNA biosensor based on magnesium oxide nanoflower and graphene oxide–gold nanoparticles hybrids coupling with enzyme signal amplification. Sens Actuators B 243:403–411

51. Thiruppathiraja C, Kamatchiammal S, Adaikkappan P, Santhosh DJ, Alagar M (2011) Specific detection of *Mycobacterium* sp. genomic DNA using dual labeled gold nanoparticle based electrochemical biosensor. Anal Biochem 417:73–79

52. Xie H, Zhang CY, Gao ZQ (2004) Amperometric detection of nucleic acid at femtomolar levels with a nucleic acid/electrochemical activator bilayer on gold electrode. Anal Chem 76:1611–1617

53. Zhang LY, Li D, Meng WL, Huang Q, Su Y, Wang LH, Song SP, Fan CH (2009) Sequence-specific DNA detection by using biocatalyzed electrochemiluminescence and non-fouling surfaces. Biosens Bioelectron 25:368–372

54. Gao ZQ, Peng YF (2011) A highly sensitive and specific biosensor for ligation- and PCR-free detection of microRNAs. Biosens Bioelectron 26:3768–3773

55. Zhang GY, Deng SY, Cai WY, Cosnier S, Zhang XJ, Shan D (2015) Magnetic zirconium hexacyanoferrate(II) nanoparticle as tracing tag for electrochemical DNA assay. Anal Chem 87:9093–9100

56. Miao XM, Wang WH, Kang TS, Liu JB, Shiu K, Leung CH, Ma D (2016) Ultrasensitive electrochemical detection of miRNA-21 by using an iridium(III) complex as catalyst. Biosens Bioelectron 86:454–458

57. Dong HF, Jin S, Ju HX, Hao KH, Xu LP, Lu HT, Zhang XJ (2012) Trace and label-free microRNA detection using oligonucleotide encapsulated silver nanoclusters as probes. Anal Chem 84:8670–8674

58. Spain E, Brennan E, McArdle H, Keyes TE, Forster RJ (2012) High sensitivity DNA detection based on regioselectively decorated electrocatalytic nanoparticles. Anal Chem 84:6471–6476

59. Chai Y, Tian DY, Wang W, Cui H (2010) A novel electrochemiluminescence strategy for ultrasensitive DNA assay using luminol functionalized gold nanoparticles multi-labeling and amplification of gold nanoparticles and biotin-streptavidin system. Chem Commun 46:7560–7562

60. Dong XY, Mi XN, Zhang L, Liang TM, Xu JJ, Chen HY (2012) DNAzyme-functionalized Pt nanoparticles/carbon nanotubes for amplified sandwich electrochemical DNA analysis. Biosens Bioelectron 38:337–341

61. Zhang K, Dong HF, Dai WH, Meng XD, Lu HT, Wu TT, Zhang XJ (2017) Fabricating Pt/Sn–In$_2$O$_3$ nanoflower with advanced oxygen reduction reaction performance for high-sensitivity microRNA electrochemical detection. Anal Chem 89:648–655

62. Kwon SJ, Bard AJ (2012) DNA analysis by application of Pt nanoparticle electrochemical amplification with single label response. J Am Chem Soc 134:10777–10779
63. Daneshpour M, Moradi LS, Izadi P, Omidfar K (2016) Femtomolar level detection of RASSF1A tumor suppressor gene methylation by electrochemical nano-genosensor based on Fe$_3$O$_4$/TMC/Au nanocomposite and PT-modified electrode. Biosens Bioelectron 77:1095–1103
64. Taton TA, Mirkin CA, Letsinger RL (2000) Scanometric DNA array detection with nanoparticle probes. Science 289:1757–1760
65. Lin L, Liu Y, Tang LH, Li JH (2011) Electrochemical DNA sensor by the assembly of graphene and DNA-conjugated gold nanoparticles with silver enhancement strategy. Analyst 136:4732–4737
66. Li H, Sun ZY, Zhong WY, Hao N, Xu DK, Chen HY (2010) Ultrasensitive electrochemical detection for DNA arrays based on silver nanoparticle aggregates. Anal Chem 82:5477–5483
67. Dong HF, Yan F, Ji HX, Wong DKY, Ju HX (2010) Quantum-dot-functionalized poly (styrene-co-acrylic acid microbeads: step-wise self-assembly, characterization, and applications for sub-femtomolar electrochemical detection of DNA hybridization. Adv Funct Mater 20:1173–1179
68. Vijian D, Chinni SV, Yin LS, Lertanantawong B, Surareungchai W (2016) Non-protein coding RNA-based genosensor with quantum dots as electrochemical labels for attomolar detection of multiple pathogens. Biosens Bioelectron 77:805–811
69. Cheng FF, He TT, Miao HH, Shi JJ, Jiang LP, Zhu JJ (2015) Electron transfer mediated electrochemical biosensor for microRNAs detection based on metal Ion functionalized titanium phosphate nanospheres at attomole level. ACS Appl Mater Interfaces 7:2979–2985

# Chapter 9
# Sandwich Assays Based on QCM, SPR, Microcantilever, and SERS Techniques for Nucleic Acid Detection

**Xiaoxia Hu and Quan Yuan**

**Abstract** Signal transducers which can read the signal toward targets are widely used for nucleic acid assay. Typically, the signal transducers based on quartz crystal microbalance (QCM), surface plasmon resonance (SPR) sensor, microcantilever, and surface-enhanced Raman scattering (SERS) play a significant role in the development of techniques for the detection of nucleic acid. The combination of these techniques with sandwich assay has received extensive attention due to the advantages of sensitivity and specificity. In this chapter, we summarized the recent development of the nucleic acid sandwich assay based on QCM, SPR sensor, microcantilever, and SERS. Additionally, the advantages and disadvantages of these sandwich assays along with the challenges and prospects are also presented, devoting to guide researches to design more of robust sandwich assays for nucleic acid assay.

**Keywords** Sandwich assay · Nucleic acid · Detection · Quartz crystal microbalance · Surface plasmon resonance · Microcantilever · Surface-enhanced Raman scattering

## 9.1 Sandwich Assays Based on QCM

QCM is a potential method to assess the surface phenomena of layers, such as antigen–antibody recognition [1, 2]. It is based on the mass change and the consequent change of piezoelectric crystals resonance frequency. Detailedly, functionalized crystal surface selectively captures the analyte, which results in the increase of the effective surface mass and the ensuing decrease in resonance frequency. Thus,

X. Hu · Q. Yuan (✉)
College of Chemistry and Molecular Sciences,
Wuhan University, Wuhan 430072, People's Republic of China
e-mail: yuanquan@whu.edu.cn

X. Hu
e-mail: xioaxiajiay@whu.edu.cn

F. Xia et al. (eds.), *Biosensors Based on Sandwich Assays*,
https://doi.org/10.1007/978-981-10-7835-4_9

the measurement of the binding event is achieved. This measurement is simple, robust and allows real-time detection. In earlier studies, Ward et al. coupled amplified mass immunosorbent assays with QCM to detect the adenosine 5′-phosphmulfate reductase and human chorionicgonadotropin. The measurement was based on the sandwich structure among anti-hCG, hCG, and anti-hCG/HRP [3].

However, it is known that the signal response of QCM-based assays in a very low target concentration is unstable, thus resulting in a low sensitivity. One way to increase the sensitivity is increasing the mass on the surface. The involvement of nanoparticles has been regarded as one of the methods. Based on this, Zhou's group developed a method for gene detection [4]. The oligonucleotide 1-functionalized QCM was hybridized with part of the target DNA 2 to form the dsDNA complex. The interaction between them resulted in the frequency decrease because of the increased surface coverage of the sensing interface. By adding the Au nanoparticle-modified oligonucleotide 3, the resulting frequency decrease was enhanced. The frequency signal was amplified because of the formation of a sandwich-type ternary complex, which consisted of an oligodeoxynucleotide immobilized on a QCM electrode, the target DNA, and an Au nanoparticle-modified oligonucleotide. Compared with the masses of the binding pair members, the mass of each nanoparticle was relatively large.

Another way to improve sensitivity is to change surface properties; Fan's group studied how single-stranded DNA (SH-ssDNA) and non-SH-ssDNA changed frequency in the presence of target DNA [5]. The result showed that the surface with thiolated SH-ssDNA attached demonstrated a sharp decrease of frequency, indicating the rapid occurrence of hybridization. However, the surface with non-SH-ssDNA attached showed no significant change in frequency.

By doing so, Su's group first developed a DNA sensor based on QCM for detection of pathogenic bacteria [6]. A thiolated single-stranded DNA (ssDNA) specific to E. coli O157:H7 eaeA gene self-assembled on the surface of QCM sensor. Then biotinylated target DNA was captured by ssDNA. The hybridization between the ssDNA probe and target DNA resulted in the mass change and consequent frequency change of the QCM. Moreover, the "mass enhancers" used in their assay was $Fe_3O_4$ nanoparticles, which amplified the frequency change. Their assay could sensitively detect 267 colony-forming units (CFU)/mL E. coli O157:H7.

Although the above assay improved the sensitivity and detection limit to some extent, it is still far from satisfactory compared to the traditional culture plating methods. Recently, Sandhyarani et al. developed a genosensor based on QCM and modified traditional sandwich assay [7]. In this work, to improve the sensitivity, gold nanoparticle was replaced by gold nanoparticle cluster (AuNPC), which conjugated with reporter probe DNA (DNA-r) for the hybridization with target DNA (DNA-t) (Fig. 9.1). The efficient immobilization of capture DNA (DNA-c) on the surface is necessary for a sensitive sensor. The surface of their sensor was modified with mercaptopropionic acid self-assembled monolayer in order to avoid the non-specific binding of the DNAs on the gold surface. The DNA sensor is based on the traditional sandwich assay. DNA-c was first immobilized on the SAM through EDC/NHS chemistry. Part of the DNA-t on the surface was complementary to DNA-c,

**Fig. 9.1** Capture DNA (DNA-c) is immobilized on the QCM surface. On treating the surface with target DNA (DNA-t), reporter probe DNA (DNA-r) conjugated gold nanoparticles cluster (DNA-r. AuNPC) can hybridize to the DNA-t, which can be monitored as a function of DNA-t concentration using frequency change of the crystal (Reprinted from Ref. [7]. Copyright 2016 Elsevier)

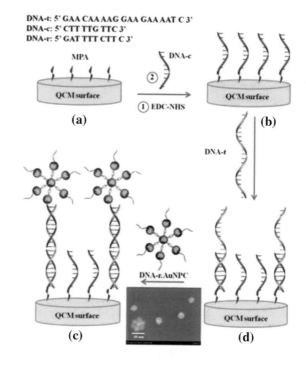

which was hybridized to the immobilized DNA-c. Then, the open part of DNA-t hybridized to the DNA conjugated with AuNPC (DNA-r.AuNPC). The hybridization of DNA-r with the DNA-t produced a large increase of mass on the surface even at ultralow concentration of DNA-t. With this method they achieved the detection of 10 aM target DNA, which enhanced the sensitivity to few orders of magnitude.

QCM was also used to monitor the multiple (re)programming of protein–DNA nanostructures. Sánchez et al. studied the binding affinity of the multi-ligand-binding flavoprotein dodecin on flavinterminated DNA monolayers based on quartz crystal microbalance with dissipation (QCM-D) measurements (Fig. 9.2) [8]. A single apododecin–flavin bond was relatively weak, and stable dodecin monolayers were formed on flavin-DNA-modified surfaces at high flavin surface coverage due to multivalent interactions between apododecin bearing six binding pockets and the surface-bound flavin-DNA ligands. If bi- or multivalent flavin ligands were adsorbed on dodecin monolayers, stable sandwich-type surface-DNA-flavin-apododecin-flavin ligand arrays were obtained. The research showed how protein-DNA nanostructures could be generated, deleted, and reprogrammed on the same surface by exploiting multivalency and the redox properties of dodecin on the same flavin-DNA-modified surface.

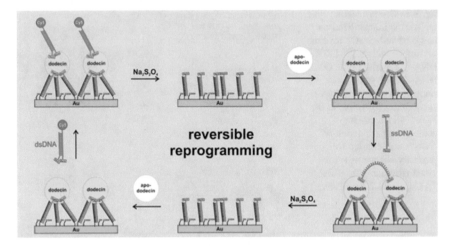

**Fig. 9.2** The binding of apododecin and bidentate flavin-DNA ligands (writing) is made of the flavin-modified dsDNA layer. After the disassembly of the surface architecture by chemical flavin reduction (erasure), dsDNA layer modified with flavin can be used for further writing and erasure cycles. Multiple reprogramming with different bidentate flavin-DNA ligands is possible (Reprinted from Ref. [8]. Copyright 2015 American Chemical Society)

## 9.2 Sandwich Assays Based on SPR

SPR sensors possess unique ability for real-time monitoring the interaction of chemical and biological analytes through measuring the refractive index changes at the SPR sensing surfaces [9–11]. In order to detect low molecular weight biological analytes (DNA) under extremely low concentration conditions, improve the specificity as well as decreasing the limit of detection, sandwich format based on nanoparticles- or/and enzymes-enhanced SPR is usually employed. Typically, the SPR sensing film is modified with capture sequences at first. Then the targeted DNA sequences are flowed onto the film and bind to the capture sequences through sequence-specific hybridization. Following, the nanoparticles tags bind to the targeted sequences through sequence-specific hybridization or/and enzymes bind to the part of the DNA-DNA/DNA-RNA sequences catalyzing the downstream reaction. Therefore, the sandwich-like structure is formed and the enhanced SPR signal can be obtained.

Gold nanoparticles (Au NPs) are one of the most commonly used signal amplification labels in nanoparticles-enhanced SPR. Duo to efficient electromagnetic coupling between Au NPs and Au sensing film, a higher sensitivity can be obtained. The detection of nucleic acids based on Au NP-enhanced SPR has attracted much attention in the past decades [12–23]. Keating et al. reported the first demonstration of DNA hybridization sensing based on Au NP-enhanced SPR [12]. The sensing film was initially functionalized with the capture sequences. Then the oligonucleotide sequence-functionalized Au NPs and the targeted DNA sequences were exposed to the SPR surface. A sandwich-like structure was formed through capture sequence-targeted DNA hybridization and targeted DNA-oligonucleotide sequences hybridization. A detection limit of 10 pM for 24-mer oligonucleotides was achieved, and such sensitivity for the target DNA has been significantly improved by more than 1000-fold compared to the unamplified binding detection method. This pioneering work demonstrated the potential of Au NP-amplified SPR for ultrasensitive detection of oligonucleotides. However, such SPR method is faced with a problem, non-specific absorption, when used for complicated samples. Misawa et al. coated a mercapto-acetic-acid (MAA) layer on the Au film to prevent undesirable DNA absorbing and optimized the length of Au NP-attached probe DNA to improve the performance of Au NP-enhanced SPR [13]. The DNA sequences with different chain lengths (15-mer, 30-mer and 60-mer) were tested, and the best sensitivity was obtained with the 30-mer DNA-functionalized Au NPs probe. With such optimized probe, target DNA within a large dynamic detection range of 1 pM to 10 mM can be detected. Zhou et al. reported that 39-mer target DNA as well as p53 cDNA can be detected with high specificity and reproducibility by combining oligonucleotide-capped gold nanoparticle with a microbore flow injection (FI) SPR setup [14]. Specifically, a carboxylated dextran film was immobilized onto the sensing film so that non-specific adsorption of oligonucleotide-capped Au NPs can be eliminated. Using this sensing strategy, 1.38 fM of 39-mer oligonucleotides and 100 fM of p53 cDNA can be detected with a remarkable sensitivity.

It is known that the size, shape, morphology of Au NPs play an important role in increasing the sensitivity of SPR [9]. Haam et al. further used a series of spherical Au NP to investigate the sensitivity enhancement properties for DNA hybridization detection [19]. In their work, Au NPs with different sizes (12–20 nm) and Au nanograting patterned sensing film were employed. The signal enhancement factor increased from 6.6-fold to 11.6-fold, and the maximum 18.2-fold was achieved with the 20 nm Au NPs coupling with the nanograting patterned sensing film. Minunni et al. used star-shaped gold nanoparticles as the nanoparticle tags and 6.9 aM of human genomic DNA can be detected [23]. Such high sensitivity was achieved thanks to the enhanced-plasmon coupling between the stars and the sensing film. In addition to DNA sequences, peptide nucleic acids (PNAs) are proposed as valuable alternatives to oligonucleotide as capture probes. Kim and co-workers utilized peptide nucleic acid as the capture probe and cationic Au nanoparticle for signal amplification by ionic interaction [18]. This method resulted in a detection limit of $58.2 \pm 1.37$ pg mL$^{-1}$. Similarly, Spoto and co-workers used PNA as the capture probe and achieved a detection limit of DNA sequences as low as 1 fM [17]. Besides, this sandwich format can remain highly sensitive in single-nucleotide mismatched recognition.

Other kinds of nanoparticles, such as hydrogel nanospheres [24], $Fe_3O_4$ nanoparticles [25, 26], and silica nanoparticles [27], have been employed for signal amplification of SPR sensors as well. Enzyme-enhanced SPR is another way to improve the sensitivity with enhancement factors compared to that of Au NP-enhanced SPR [28–33]. Gao et al. reported a signal amplification strategy with the assistance of DNA-guided polyaniline deposition [30]. The target DNA hybridized with the PNA probe first and DNA-templated polyaniline deposition was followed in the presence of $H_2O_2$ and horseradish peroxidase. The in situ polymer chain growth along DNA strands contributed to the sensitivity improvement and 50-fold improvement of the limit of detection was achieved. Corn and co-workers used RNase H that can specifically digest RNA oligonucleotides in RNA-DNA heteroduplex to gain significantly enzymatic amplification [28]. When the target ssDNA binding to the ssRNA capture probe immobilized on the sensing microarrays to form a heteroduplex, RNase H can digest the ssRNA and the target ssDNA can be released. The released DNA was further hybridized with another ssRNA and induced another enzymatic hydrolysis. Such repeatable cycle resulted in enzymatic amplification sensitivity by 6 orders of magnitude. Therefore, DNA targets with concentration down to 10 fM can be detected.

By combing signal enhancement based on the enzymatic amplification and nanoparticles, the sensitivity can be further improved [11]. In 2006, Corn et al. put forward a novel approach to detect multiple microRNAs by combining a surface enzyme reaction with nanoparticle-amplified SPR imaging (SPRI) [34]. In their work, three kinds of miRNAs (miR-16, miR-122b, miR-23b) obtained from mouse liver tissue were detected and locked nucleic acids (LNAs) were specifically used as the capture probe for these miRNAs. The proposed method includes three steps. The LNAs were initially immobilized onto a microarray format and the targeted miRNAs hybridized with the complementary LNAs. Then poly(A) polymerase was added into the array to form a poly(A) tail on the miRNAs. Finally, the solution of $T_{30}$ functionalized Au NPs was flowed onto the sensing surface and hybridized with poly(A) tails. Through such ultrasensitive NPs-amplified SPRI methodology, a detection limit of 10 fM was gained. In 2011, Corn et al. proposed a signal amplification strategy by coupling a surface RNA transcription reaction to nanoparticle-enhanced SPRI (Fig. 9.3) [35]. In their design, two kinds of adjacent microarrays, one called generator for RNA transcription and the other called detector for Au NP-enhanced detection of the transcribed RNA, were employed. The capture DNA sequences containing T7 promoter sequence was modified on the generator microarray and then target DNA sequences hybridize with the capture DNA to form dsDNA templates. In the presence of T7 RNA polymeras, numerous ssRNA copies were synthesized. Such transcribed ssRNA can diffuse to the detector microarray and be captured by the second ssDNA immobilized on the detector. ssDNA-functionalized AuNPs were added into the system and absorbed on the detector through hybridizing with the transcribed ssRNA. This dual amplification method can be used to detect ssDNA down to 1 fM.

In addition to the methods mentioned above, there are other types of amplification strategy. Szunerits and co-workers demonstrated that DNA in the attomolar concentration range can be detected with SPR by non-covalent coating graphene layers on gold sensing film [36]. Recently, Wang et al. proposed a multiple signal amplification strategy for miRNA detection (Fig. 9.4) [37]. In their strategy, a hairpin probe was employed as the capture probe and immobilized on the Au film first. Then target miRNA was hybridized with the hairpin probe; therefore, the stem-loop was unfolded and the DNA-functionalized Au NPs can hybridize with

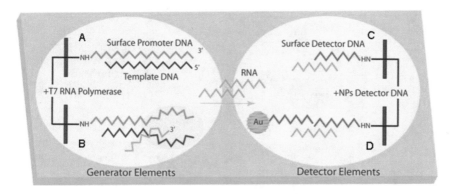

**Fig. 9.3** Schematic illustration of the signal amplification strategy. **a** On the generator elements a surface promoter DNA is covalently attached to the gold surface and then hybridizes with template DNA from solution. **b** An in situ surface RNA polymerase reaction is used to transcribe numerous ssRNA copies. **c** The ssRNA is base-paired with the surface detector DNA. **d** DNA-modified AuNPs can bind to the detector elements via RNA hybridization and the ssRNA is detected with nanoparticle-enhanced SPRI (Reprinted from Ref. [35]. Copyright 2011 American Chemical Society)

the terminus of the unfold hairpin. Subsequently, two kinds of report DNA sequences were added into the above system and DNA supersandwich structure was formed. Finally, numerous positively charged Ag NPs were added and absorbed onto the long-range DNA supersandwich. Through such strategy, 0.6 fM of miRNA-21 can be detected and single-base mismatch can be sensitively recognized. The multiple signal amplification was achieved through three main factors: (1) enhanced-electronic coupling between localized plasmon of the Au NPs and surface plasmon of the sensing film; (2) enhanced-refractive index of the medium induced by DNA supersandwich structure; (3) enhanced-electronic coupling between localized plasmon of the Ag NPs and surface plasmon of the sensing film.

## 9.3  Sandwich Assays Based on the Microcantilever

Microcantilever has recently been emerging great attention in the field of chemical, physical, and biological detection [23–26, 38–43]. The fundamental principle for microcantilever-based assay is that the adsorption of molecular on one cantilever surface can change the surface stress that causes the mechanical bending deflection motion of the cantilever [44–47]. By selecting adsorbed probe that can recognize the specific molecular, it has the possibility to detect various targeted molecules. Like QCM and SPR, microcantilever can directly transfer the molecular recognition into nanomechanics and does not need the labeling of targets. Serving as an upcoming sensing technique, microcantilever exhibits many advantages such as high sensitivity, potential low cost, and faster response time. It has broad application in chemical, physical, and biological detection. For instance, Gerber's group developed

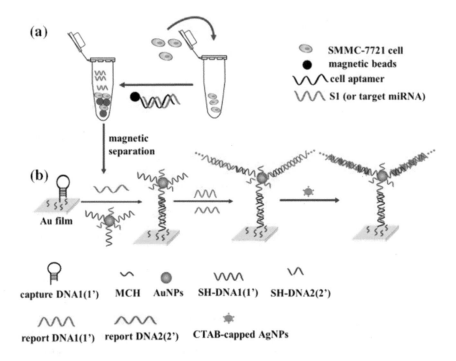

**Fig. 9.4** Schema of SPR biosensor. **a** The SMMC-7721 cell aptamer-modified magnetic beads were hybridized with the complementary sequence, and then were incubated with cells. S1 probe was released through magnetic separation, and the S1 probe could act as the target. **b** The stem-loop structure was unfold in the presence of S1 probe, and then DNA-linked gold nanoparticles was hybridized with S1 probe. DNA supersandwich structure was formed upon introducing report DNA1 and report DNA2. Numerous positively charged silver nanoparticles (AgNPs) were bound to the DNA supersandwich, resulting in a increase of resonance angle shift (Reprinted with permission from Ref. [37]. Copyright 2017 Elsevier)

a method based on microcantilever arrays to detect the mutation in total RNA samples extracted from melanoma cells [40]. The BEAF-specific oligonucleotide probe was linked on one cantilever surface. The target DNA or RNA containing the matching sequence and the other non-related sequences was injected to the micro-cantilever arrays. On hybridization, only the probe cantilever bended, but no binding occurred on the reference cantilever, thus giving a differential deflection (Fig. 9.5). They detected the mutant BRAF at a concentration of 0.5 nM in a 50-fold excess of wild-type sequence. This method had an ability to distinguish melanoma cells with mutation BRAF using the RNA concentration as low as 20 ng $\mu L^{-1}$, without using PCR amplification. Gerber et al. reported microfabricated cantilevers for DNA hybridization assay [23]. They immobilized a selection of receptor molecules on one cantilever and then detected the mechanical bending induced by the ligand binding. The differential deflection of the microcantilevers provided a true molecular recognition signal, and they monitored the bending of each cantilever in real time by

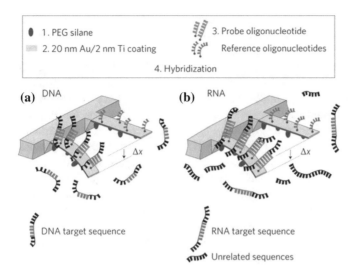

**Fig. 9.5** Principle of microcantilever array for the detection of a BRAF mutation in total RNA. The cantilevers were modified with a probe oligonucleotide (in red) and a reference oligonucleotide (in blue). On treating with target DNA or RNA, only the probe cantilever bends, and then the differential deflection $\Delta x$ increased (Reprinted with permission from Ref. [40]. Copyright 2013 Nature Publishing Group)

investigating changes in the optical beam deflection. They have shown that a single-base mismatch between two 12-mer oligonucleotides could be clearly detected. In the opposite way, the thermal dehybridization of double-stranded DNA on the cantilever surface was investigated by Majumdar's group [46]. They used heat to separate the double-stranded DNA into two single strands. The dehybridization of double helix on one microcantilever beam could lead to one complementary DNA diffused away from the other DNA strand. Therefore, the change of surface stress was observed. They have successfully distinguished the changes in the melting temperature of double-strand DNA on the basis of salt concentration and oligomer length. Interestingly, McKendry et al. reported that the force generated by an i-motif conformational change could be probed using the micromechanical cantilever arrays coated with a non-specific sequence of DNA [48].

Improving detection sensitivity is of significance for clinical diagnosis and various genome projects. In the previous study, Dravid's group has developed a sandwich-based microcantilever for DNA detection using gold nanoparticle-modified probes [45]. The capture DNA was firstly linked on the cantilever, and then the target DNA was hybridized with the capture DNA. The gold nanoparticle-labeled DNA strand was integrated into the capture DNA–target DNA complex through the complementary interactions. Gold nanoparticles served as a nucleating agent for the growth of silver, which could lead to a detectable frequency shift due to the increasing of the mass of the microcantilever. The core strategy of this idea is that the DNA hybridization can cause the mass change of a microfabricated cantilever, and the signal can be amplified by gold nanoparticle-catalyzed nucleation of silver. They can detect the target DNA concentration down to 0.05 nM. In addition, a single-base mismatch can be discriminated.

According to recent studies, dynamic-mode millimeter-sized cantilevers can detect the oligonucleotides at extremely lower concentration in comparison to the static-mode microcantilevers. Kim et al. have developed a silica nanoparticle-enhanced dynamic microcantilever biosensor for Hepatitis B Virus (HBV) DNA detection (Fig. 9.6) [49]. The capture DNA was immobilized on the microcantilever surface. Then, the HBV target DNA could hybridize with the capture DNA, and the silica nanoparticle-labeled probe DNA was conjugated with this capture DNA-probe DNA complex. To make the silica nanoparticle efficiency to enhance the detection sensitivity, they optimized the size of the silica nanoparticle and the dimension of the microcantilever. Without nanoparticle enhancement, the HBV target DNA was detected up to pM level. When the silica nanoparticle-based signal amplification process was applied, they could detect the concentration of HBV target DNA at fM level.

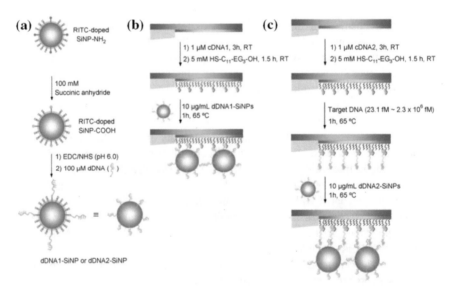

**Fig. 9.6** Schematic illustration of DNA assay using the silica nanoparticle (SiNP)-enhanced microcantilever. **a** SiNP was modified with detection DNA. **b** dDNA was hybridized with cDNA on the microcantilever to optimize conditions of the SiNPs-enhanced DNA detection. **c** In the presence of the target DNA, the nanoparticle-based sandwich assay was formed. The concentration for the target DNA was monitored by the resonant frequency shifts (Reprinted with permission from Ref. [49]. Copyright 2009 Elsevier)

## 9.4   Sandwich Assays Based on SERS

SESR is a surface-sensitive technique that can enhance the Raman scattering by absorbing molecules on rough metal surfaces or by nanostructures [50]. The phenomenon of SERS is commonly explained by combining an electromagnetic mechanism (reflecting the surface electron movement in the substrate) and a chemical mechanism (relating to charge transfer between substrate and target molecules) [42, 51–53]. Mirkin's group reported a sandwich assay strategy based on the Raman spectroscopic fingerprints for multiple DNA and RNA detection [38]. They designed the nanoparticle probes by labeling the gold nanoparticles with specific oligonucleotides and Raman dyes. The silver-coated gold nanoparticles were employed as a surface-enhanced Raman scattering promotor. By integrating with SERS spectroscopy, the Raman spectroscopic fingerprint could be identified by scanning Raman spectroscopy. A series of Raman scattering response could be obtained. Therefore, a large number of oligonucleotides with different sequences could be detected. The detection limit of this strategy was 20 fM.

To make the nucleic acid detection more sensitive and stable, Yang et al. found that the semiconductor nanoparticles which have Raman signal were more applicable for achieving the high sensitivity detection than the dye molecules [54]. Since ZnO quantum dots (QDs) are capable of transferring electrons to gold nanoparticles, ZnO/Au nanocomposites can achieve the electromagnetic-field enhancement. They functionalized the thiol-oligonucleotides with ZnO/Au nanocomposites as Raman labels. The capture DNA was firstly immobilized on the gold film, and the target DNA and the ZnO/Au functionalized probe DNA could be hybridized with the capture DNA to form a stable sandwich structure. With a strong resonance Raman scattering signal output, the target oligonucleotide strand could be detected with extraordinary sensitivity and selectivity (Fig. 9.7).

A SERS "hot spot" is predicted to be created by forming a junction between nanoparticles and smooth surface. Reich and Moskovists developed a versatile SERS biosensor by assembling the probe DNA-tethered Ag nanoparticles to the smooth Ag surface due to the hybridization of the target DNA with the capture DNA and probe DNA [48]. The "hot spot" was thus created to enhance the Raman signals. The intense and reliable SERS signals could be obtained at near-single particle level. Furthermore, it was indicated that the decrease of the distance between the nanoparticle and Raman molecule can enhance the Raman signal. Based on this principle, by forming a SERS "hot spot" between the nanoparticles or between the nanoparticles and surface, the sandwiched structure-based DNA biosensor has been developed for HIV-1 DNA detection by Liu's group [55]. The target DNA triggered the formation of sandwich structure of capture DNA–target DNA–probe DNA. Then, the probe DNA was further recognized by another Raman tag-labeled probe DNA. The multi-metal-molecule-metal-sandwiched structure was finally constructed, which created a SERS "hot spot" and further decreased the distance between the gold nanoparticles and Raman tags. Therefore, the Raman signal has been largely enhanced and this sandwich-based platform could detect

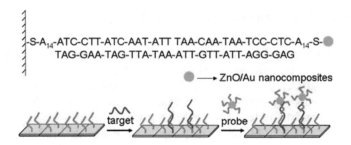

**Fig. 9.7** Capture DNA was attached to the Au film. In the presence of target DNA, ZnO/Au nanocomposites-based sandwich assay was formed, and the ZnO/Au nanocomposites were used as Raman labels (Reprinted with permission from Ref. [54]. Copyright 2008 American Chemical Society)

DNA concentration down to $10^{-19}$ M with the capability for distinguishing the single-base mismatch.

To control the assembly of nanoparticles to turn on the Raman signal enhancement in a reproducible manner, Faulds developed a DNA-based assembly process to control the enhanced Raman scattering [56]. They applied the DNA hybridization to control the enhancement of Raman scattering from Ag nanoparticles labeled with a Raman dye. Here, two mechanisms were involved to be responsible for the signal enhancement, which were chemical and electromagnetic. A monolayer dye was firstly modified on the Ag nanoparticles; two different oligonucleotide probe sequences were then linked on the surface of Ag nanoparticles, respectively. The target DNA sequence finally integrated with these two different Raman dye-labeled Ag nanoparticles to form a controlled assembly. The signal was only capable of being obtained when the target recognition event of DNA hybridization taken place. Similar to the strategy reported by Faulds, Graham et al. precisely controlled the assembly of dye-coded and oligonucleotide-modified Ag nanoparticle conjugates [49]. This strategy could discriminate the single mismatched base in an unmodified target oligonucleotide.

To achieve the simultaneous multiple detection of nucleic acid, Song and Wen developed a SERS-based sandwich method using the mixed DNA-functionalized Ag nanoparticles [57]. Three kinds of probe DNA strands were co-assembled at the surface of the Ag nanoparticles at equal molar ratios to form conjugate 1. In addition, they prepared three kinds of stable Ag nanoparticle-oligonucleotide conjugates based on the Raman dyes and triple-cyclic disulfide-modified DNA strands. The targeted DNA could be hybridized with the conjugate 1 and the corresponding probe conjugate 2, and the electromagnetic enhancement of Raman dyes labeled on these nanoparticles is therefore enhanced (Fig. 9.8). Thus, the specific detection of multiple target DNA could be achieved.

Wang showed that enriching the target-mediated Raman tags aggregation is a promising strategy to improve the detection sensitivity [58]. They attached the probe DNA to the silica-Ag nanoparticles composite and labeled the magnetic

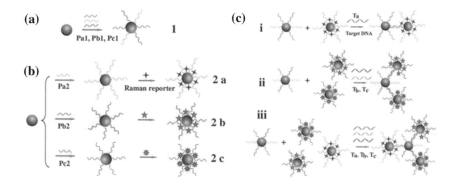

**Fig. 9.8** **a** and **b** A scheme of the preparation of the mixed DNA-modified AgNPs 1 and Raman dye and DNA-modified AgNPs 2, respectively. **c** The SERS-based sandwich detection system for one (I), two (II), three (III) target DNA detection (Reprinted with permission from Ref. [57]. Copyright 2011 The Royal Society of Chemistry)

nanospheres with the capture DNA. The target DNA was first allowed to hybridize with the SERS tag-labeled probe DNA. With the addition of the capture DNA-linked magnetic nanospheres, it could be integrated with the above composite. When the external was applied to the reaction solution, the nanocomposites were deposited together and then could be separated and analyzed by SERS (Fig. 9.9). They realized the quantitative detection of target DNA in the range of

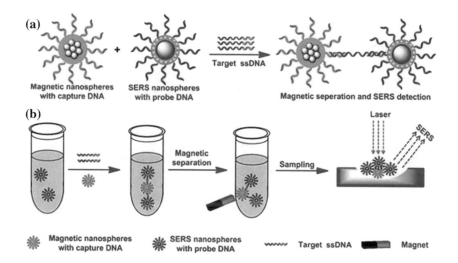

**Fig. 9.9** **a** Schematic representation of a three-component-based sandwich assay in the presence of target ssDNA. **b** Scheme of the experimental procedure for the detection of target ssDNA (Reprinted with permission from Ref. [58]. Copyright 2013 American Chemical Society)

10 nM to 10 pM. In addition, they achieved the multiplexed detection of up to three different target DNAs.

## 9.5 Conclusion

The analysis of nucleic acids is important in the research of their fundamental functions. Developing sensitive, simple, and specific detection techniques for DNA detection is of great significance for the development of molecular diagnostics. In this chapter, we have described the recent developments in the detection of nucleic acid by the sandwich assays based on QCM, SPR sensor, microcantilever, and SERS. Sandhyarani et al. achieved the ultrasensitive detection of DNA by using the QCM-based sandwich assay. For SPR-based sandwich assay, various nanoparticles such as AuNPs, hydrogel nanospheres, $Fe_3O_4$ nanoparticles, and silica nanoparticles have been used as signal amplification labels for signal amplification of SPR sensors. This amplification method can be used to detect ssDNA down to 1 fM. In addition, by applying the silica nanoparticle-based signal amplification process to the microcantilevers-based sandwich assay, Kim et al. achieved the detection of HBV target DNA at fM level. The simultaneous multiple detection of nucleic acid was achieved by Song and Wen by using the SERS-based sandwich assay. The large achievement of the sandwich assay was realized in recent decades, but there are still some problems such as complex process, non-specific adsorption, and stability need to be solved. This relies on the further development of sandwich assay to resist these problems.

## References

1. Henne WA, Doorneweerd DD, Lee J, Low PS, Savran C (2006) Detection of folate binding protein with enhanced sensitivity using a functionalized quartz crystal microbalance sensor. Anal Chem 78:4880–4884
2. Uludag Y, Tothill IE (2012) Cancer biomarker detection in serum samples using surface plasmon resonance and quartz crystal microbalance sensors with nanoparticle signal amplification. Anal Chem 84:5898–5904
3. Ebersole RC, Ward MD (1988) Amplified mass immunosorbent-assay with a quartz crystal microbalance. J Am Chem Soc 110:8623–8628
4. Zhou XC, O'Shea SJ, Li SFY (2000) Amplified microgravimetric gene sensor using Au nanoparticle modified oligonucleotides. Chem Commun 953–954
5. Lao RJ, Song SP, Wu HP, Wang LH, Zhang ZZ, He L, Fan CH (2005) Electrochemical interrogation of DNA monolayers on gold surfaces. Anal Chem 77:6475–6480
6. Mao XL, Yang LJ, Su XL, Li YB (2006) A nanoparticle amplification based quartz crystal microbalance DNA sensor for detection of *Escherichia coli* O157:H7. Biosens Bioelectron 21:1178–1185
7. Rasheed PA, Sandhyarani N (2016) Quartz crystal microbalance genosensor for sequence specific detection of attomolar DNA targets. Anal Chim Acta 905:134–139

8. Sanchez CG, Su Q, Schonherr H, Grininger M, Noll G (2015) Multi-ligand-binding flavoprotein dodecin as a key element for reversible surface modification in nano-biotechnology. ACS Nano 9:3491–3500
9. Zeng SW, Baillargeat D, Ho HP, Yong KT (2014) Nanomaterials enhanced surface plasmon resonance for biological and chemical sensing applications. Chem Soc Rev 43:3426–3452
10. Ermini ML, Mariani S, Scarano S, Minunni M (2014) Bioanalytical approaches for the detection of single nucleotide polymorphisms by Surface Plasmon Resonance biosensors. Biosens Bioelectron 61:28–37
11. Sipova H, Homola J (2013) Surface plasmon resonance sensing of nucleic acids: a review. Anal Chim Acta 773:9–23
12. He L, Musick MD, Nicewarner SR, Salinas FG, Benkovic SJ, Natan MJ, Keating CD (2000) Colloidal Au-enhanced surface plasmon resonance for ultrasensitive detection of DNA hybridization. J Am Chem Soc 122:9071–9077
13. Hayashida M, Yamaguchi A, Misawa H (2005) High sensitivity and large dynamic range surface plasmon resonance sensing for DNA hybridization using Au-nanoparticle-attached probe DNA. Jpn J Appl Phys Part 2—Lett Express Lett 44:1544–1546
14. Yao X, Li X, Toledo F, Zurita-Lopez C, Gutova M, Momand J, Zhou FM (2006) Sub-attomole oligonucleotide and p53 cDNA determinations via a high-resolution surface plasmon resonance combined with oligonucleotide-capped gold nanoparticle signal amplification. Anal Biochem 354:220–228
15. Yang XH, Wang Q, Wang KM, Tan WH, Li HM (2007) Enhanced surface plasmon resonance with the modified catalytic growth of Au nanoparticles. Biosens Bioelectron 22:1106–1110
16. Wark AW, Lee HJ, Qavi AJ, Corn RM (2007) Nanoparticle-enhanced diffraction gratings for ultrasensitive surface plasmon biosensing. Anal Chem 79:6697–6701
17. D'Agata R, Corradini R, Grasso G, Marchelli R, Spoto G (2008) Ultrasensitive detection of DNA by PNA and nanoparticle-enhanced surface plasmon resonance imaging. ChemBioChem 9:2067–2070
18. Joung HA, Lee NR, Lee SK, Ahn J, Shin YB, Choi HS, Lee CS, Kim S, Kim MG (2008) High sensitivity detection of 16s rRNA using peptide nucleic acid probes and a surface plasmon resonance biosensor. Anal Chim Acta 630:168–173
19. Moon S, Kim DJ, Kim K, Kim D, Lee H, Lee K, Haam S (2010) Surface-enhanced plasmon resonance detection of nanoparticle-conjugated DNA hybridization. Appl Optics 49:484–491
20. D'Agata R, Breveglieri G, Zanoli LM, Borgatti M, Spoto G, Gambari R (2011) Direct detection of point mutations in nonamplified human genomic DNA. Anal Chem 83:8711–8717
21. Hong X, Hall EAH (2012) Contribution of gold nanoparticles to the signal amplification in surface plasmon resonance. Analyst 137:4712–4719
22. Gu Y, Tan YJ, Wang CY, Nie JL, Yu JR, Li YH (2012) A surface plasmon resonance sensor platform coupled with gold nanoparticle probes for unpurified nucleic acids detection. Anal Lett 45:2210–2220
23. Mariani S, Scarano S, Spadavecchia J, Minunni M (2015) A reusable optical biosensor for the ultrasensitive and selective detection of unamplified human genomic DNA with gold nanostars. Biosens Bioelectron 74:981–988
24. Okumura A, Sato Y, Kyo M, Kawaguchi H (2005) Point mutation detection with the sandwich method employing hydrogel nanospheres by the surface plasmon resonance imaging technique. Anal Biochem 339:328–337
25. Mousavi MZ, Chen HY, Wu SH, Peng SW, Lee KL, Wei PK, Cheng JY (2013) Magnetic nanoparticle-enhanced SPR on gold nanoslits for ultra-sensitive, label-free detection of nucleic acid biomarkers. Analyst 138:2740–2748
26. Mousavi MZ, Chen HY, Lee KL, Lin H, Chen HH, Lin YF, Wong CS, Li HF, Wei PK, Cheng JY (2015) Urinary micro-RNA biomarker detection using capped gold nanoslit SPR in a microfluidic chip. Analyst 140:4097–4104

27. Zhou WJ, Halpern AR, Seefeld TH, Corn RM (2012) Near infrared surface plasmon resonance phase imaging and nanoparticle-enhanced surface plasmon resonance phase imaging for ultrasensitive protein and DNA biosensing with oligonucleotide and aptamer microarrays. Anal Chem 84:440–445

28. Goodrich TT, Lee HJ, Corn RM (2004) Enzymatically amplified surface plasmon resonance imaging method using RNase H and RNA microarrays for the ultrasensitive detection of nucleic acids. Anal Chem 76:6173–6178

29. Zeng DM, Wang JX, Yin LJ, Zhang YT, Zhang Y, Zhou FM (2007) Sequence-specific analysis of oligodeoxynucleotides by precipitate-amplified surface plasmon resonance measurements. Front Biosci 12:5117–5123

30. Su XD, Teh HF, Aung KMM, Zong Y, Gao ZQ (2008) Femtomol SPR detection of DNA-PNA hybridization with the assistance of DNA-guided polyaniline deposition. Biosens Bioelectron 23:1715–1720

31. Seefeld TH, Zhou WJ, Corn RM (2011) Rapid microarray detection of DNA and proteins in microliter volumes with surface plasmon resonance imaging measurements. Langmuir 27:6534–6540

32. Fasoli JB, Corn RM (2015) Surface enzyme chemistries for ultrasensitive microarray biosensing with SPR imaging. Langmuir 31:9527–9536

33. Li YA, Wark AW, Lee HJ, Corn RM (2006) Single-nucleotide polymorphism genotyping by nanoparticle-enhanced surface plasmon resonance imaging measurements of surface ligation reactions. Anal Chem 78:3158–3164

34. Fang SP, Lee HJ, Wark AW, Corn RM (2006) Attomole microarray detection of MicroRNAs by nanoparticle-amplified SPR imaging measurements of surface polyadenylation reactions. J Am Chem Soc 128:14044–14046

35. Sendroiu IE, Gifford LK, Luptak A, Corn RM (2011) Ultrasensitive DNA microarray biosensing via in situ RNA transcription-based amplification and nanoparticle-enhanced SPR imaging. J Am Chem Soc 133:4271–4273

36. Zagorodko O, Spadavecchia J, Serrano AY, Larroulet I, Pesquera A, Zurutuza A, Boukherroub R, Szunerits S (2014) Highly sensitive detection of DNA hybridization on commercialized graphene-coated surface plasmon resonance interfaces. Anal Chem 86:11211–11216

37. Liu RJ, Wang Q, Li Q, Yang XH, Wang KM, Nie WY (2017) Surface plasmon resonance biosensor for sensitive detection of microRNA and cancer cell using multiple signal amplification strategy. Biosens Bioelectron 87:433–438

38. Cao YWC, Jin RC, Mirkin CA (2002) Nanoparticles with Raman spectroscopic fingerprints for DNA and RNA detection. Science 297:1536–1540

39. Ghosh S, Mishra S, Mukhopadhyay R (2014) Enhancing sensitivity in a piezoresistive cantilever-based label-free DNA detection assay using ssPNA sensor probes. J Mat Chem B 2:960–970

40. Huber F, Lang HP, Backmann N, Rimoldi D, Gerber C (2013) Direct detection of a BRAF mutation in total RNA from melanoma cells using cantilever arrays. Nat Nanotechnol 8:125–129

41. Mertens J, Rogero C, Calleja M, Ramos D, Martin-Gago JA, Briones C, Tamayo J (2008) Label-free detection of DNA hybridization based on hydration-induced tension in nucleic acid films. Nat Nanotechnol 3:301–307

42. Zhu R, Howorka S, Proll J, Kienberger F, Preiner J, Hesse J, Ebner A, Pastushenko VP, Gruber HJ, Hinterdorfer P (2010) Nanomechanical recognition measurements of individual DNA molecules reveal epigenetic methylation patterns. Nat Nanotechnol 5:788–791

43. Zheng S, Choi JH, Lee SM, Hwang KS, Kim SK, Kim TS (2011) Analysis of DNA hybridization regarding the conformation of molecular layer with piezoelectric microcantilevers. Lab Chip 11:63–69

44. McKendry R, Zhang JY, Arntz Y, Strunz T, Hegner M, Lang HP, Baller MK, Certa U, Meyer E, Guntherodt HJ, Gerber C (2002) Multiple label-free biodetection and quantitative

DNA-binding assays on a nanomechanical cantilever array. Proc Natl Acad Sci U S A 99:9783–9788

45. Su M, Li SU, Dravid VP (2003) Microcantilever resonance-based DNA detection with nanoparticle probes. Appl Phys Lett 82:3562–3564

46. Wu GH, Ji HF, Hansen K, Thundat T, Datar R, Cote R, Hagan MF, Chakraborty AK, Majumdar A (2001) Origin of nanomechanical cantilever motion generated from biomolecular interactions. Proc Natl Acad Sci U S A 98:1560–1564

47. Lee SM, Hwang KS, Yoon HJ, Yoon DS, Kim SK, Lee YS, Kim TS (2009) Sensitivity enhancement of a dynamic mode microcantilever by stress inducer and mass inducer to detect PSA at low picogram levels. Lab Chip 9:2683–2690

48. Shu WM, Liu DS, Watari M, Riener CK, Strunz T, Welland ME, Balasubramanian S, McKendry RA (2005) DNA molecular motor driven micromechanical cantilever arrays. J Am Chem Soc 127:17054–17060

49. Cha BH, Lee SM, Park JC, Hwang KS, Kim SK, Lee YS, Ju BK, Kim TS (2009) Detection of Hepatitis B Virus (HBV) DNA at femtomolar concentrations using a silica nanoparticle-enhanced microcantilever sensor. Biosens Bioelectron 25:130–135

50. Xu XB, Li HF, Hasan D, Ruoff RS, Wang AX, Fan DL (2013) Near-field enhanced plasmonic-magnetic bifunctional nanotubes for single cell bioanalysis. Adv Funct Mater 23:4332–4338

51. Banholzer MJ, Millstone JE, Qin LD, Mirkin CA (2008) Rationally designed nanostructures for surface-enhanced Raman spectroscopy. Chem Soc Rev 37:885–897

52. Larmour IA, Graham D (2011) Surface enhanced optical spectroscopies for bioanalysis. Analyst 136:3831–3853

53. Prado E, Daugey N, Plumet S, Servant L, Lecomte S (2011) Quantitative label-free RNA detection using surface-enhanced Raman spectroscopy. Chem Commun 47:7425–7427

54. Liu YC, Zhong MY, Shan GY, Li YJ, Huang BQ, Yang GL (2008) Biocompatible ZnO/Au nanocomposites for ultrasensitive DNA detection using resonance Raman scattering. J Phys Chem B 112:6484–6489

55. Hu J, Zheng PC, Jiang JH, Shen GL, Yu RQ, Liu GK (2010) Sub-attomolar HIV-1 DNA detection using surface-enhanced Raman spectroscopy. Analyst 135:1084–1089

56. Graham D, Thompson DG, Smith WE, Faulds K (2008) Control of enhanced Raman scattering using a DNA-based assembly process of dye-coded nanoparticles. Nat Nanotechnol 3:548–551

57. Zhang ZL, Wen YQ, Ma Y, Luo J, Jiang L, Song YL (2011) Mixed DNA-functionalized nanoparticle probes for surface-enhanced Raman scattering-based multiplex DNA detection. Chem Commun 47:7407–7409

58. Li JM, Ma WF, You LJ, Guo J, Hu J, Wang CC (2013) Highly sensitive detection of target ssDNA based on SERS liquid chip using suspended magnetic nanospheres as capturing substrates. Langmuir 29:6147–6155

59. Sipova H, Zhang SL, Dudley AM, Galas D, Wang K, Homola J (2010) Surface plasmon resonance biosensor for rapid label-free detection of microribonucleic acid at subfemtomole level. Anal Chem 82:10110–10115

60. Liu JY, Tian SJ, Tiefenauer L, Nielsen PE, Knoll W (2005) Simultaneously amplified electrochemical and surface plasmon optical detection of DNA hybridization based on ferrocene-streptavidin conjugates. Anal Chem 77:2756–2761

61. Ding XJ, Yan YR, Li SQ, Zhang Y, Cheng W, Cheng Q, Ding SJ (2015) Surface plasmon resonance biosensor for highly sensitive detection of microRNA based on DNA super-sandwich assemblies and streptavidin signal amplification. Anal Chim Acta 874:59–65

# Chapter 10
# Sandwich Assays for Small Molecule and Ion Detection

Yu Dai, Xiaojin Zhang and Fan Xia

**Abstract** Small molecules and ions play a critical role in biological and environmental systems. The detection of small molecules and ions is a significantly important issue and still a challenge in analytical chemistry. In the past decade, large attention has been paid to the detection of small molecules and ions based on sandwich assays due to their high sensitivity and selectivity. In this chapter, we summarize some sandwich assays for the detection of small molecules and ions that were proposed in recent years. The detection techniques afforded in the sandwich assays for the detection of small molecules and ions include electrochemical method, electrochemiluminescence method, fluorescence method, colorimetric method, and some other methods such as surface plasmon resonance (SPR), surface-enhanced Raman scattering (SERS), and quartz crystal microbalance (QCM).

**Keywords** Small molecules and ions · Sandwich assays · Electrochemical method
Electrochemiluminescence method · Fluorescence method · Colorimetric method

The original version of this chapter was revised: Foreword has been included and authors' affiliations have been updated. The erratum to this chapter is available at https://doi.org/10.1007/978-981-10-7835-4_13

Y. Dai · X. Zhang (✉) · F. Xia
Engineering Research Center of Nano-Geomaterials of Ministry of Education,
Faculty of Materials Science and Chemistry, China University of Geosciences,
Wuhan 430074, People's Republic of China
e-mail: zhangxj@cug.edu.cn

Y. Dai
e-mail: yudai@cug.edu.cn

F. Xia
e-mail: xiafan@cug.edu.cn; xiafan@hust.edu.cn

F. Xia
Hubei Key Laboratory of Bioinorganic Chemistry & Materia Medica,
School of Chemistry and Chemical Engineering, Huazhong University of Science
and Technology, Wuhan 430074, People's Republic of China

© Springer Nature Singapore Pte Ltd. 2018
F. Xia et al. (eds.), *Biosensors Based on Sandwich Assays*,
https://doi.org/10.1007/978-981-10-7835-4_10

167

## 10.1 Introduction

Small molecules are usually at very low concentrations. Some are as poisonous substances in living cells or organisms such as toxins [1], while others play a very critical role in the functions in biological systems such as adenosine triphosphate (ATP), which significantly contributes to cell signaling, cell locomotion, metabolism, and active transport [2]. Therefore, it is an important concern to rapidly and accurately detect small molecules, which is still a challenge in analytical chemistry [3]. The typical method for the detection of small molecules is high-performance liquid chromatography (HPLC) with UV and/or fluorescence detection. This method provides sensitive and specific detection of small molecules but has some limitations such as limited sample dose, time-consuming operation, and complex analysis procedure. Biosensors used for the detection of small molecules have been rapidly developed in recent years due to their rapid detection, convenient operation, and simple protocol [4]. Among them, numerous efforts focusing on the sandwich assays for the detection of small molecules have also been made in the past decade.

Metal ions play an important role in biological and environmental systems [5]. Most metal ions, for example, mercury ($Hg^{2+}$), lead ($Pb^{2+}$), silver ($Ag^+$), copper ($Cu^{2+}$) ions, are toxic and cause various diseases if they are accumulated in human bodies [6]. In addition, metal ion contamination is also a serious environmental problem in developing countries [7]. Accordingly, the detection and concentration monitoring of metal ions is of critical importance in biological and environmental systems [8]. However, it has always been a challenge to detect metal ions with biosensor in analytical chemistry [9]. Fortunately, numerous efforts have been reported to develop rapid and accurate methods for the detection of metal ions. Among them, the methods based on the sandwich assays have shown great potential in the detection of metal ions due to their high sensitivity and selectivity, low cost, and easy operation.

The sandwich assays employed in biosensors have achieved great success in the detection of proteins and nucleic acids mentioned in the above chapters. In this chapter, we will highlight the detection of small molecules and ions using the sandwich assays that are developed in the past decade. It is well known that there are various methods that have been employed in the sandwich assays for the detection of small molecules and ions, such as electrochemical method, electrochemiluminescence method, fluorescence method, colorimetric method, and other methods. Next, we will introduce them from small molecules detection and ions detection through a series of examples.

## 10.2   Sandwich Assays for Small Molecule Detection

### 10.2.1   Electrochemical Sandwich Assays

An electrochemical sandwich assay based on single aptamer sequences immobilized on the surface of gold electrode to detect cocaine and ATP was demonstrated by Plaxco et al., as shown in Fig. 10.1 [10]. The mechanism was exploited that the target molecule (cocaine or ATP) bound the aptamers, leading to significant increase of faradaic current from methylene blue monitored via voltammetry. The result showed that an approximately 600 and 400% increase in faradaic current was observed for the detection of cocaine and ATP, respectively. Gold nanoparticle (AuNP)-attached aptamers were introduced into this assay to improve the binding affinity of anti-cocaine split aptamer pairs by up to 66-fold, resulting in 1000-fold lower reporter probe concentrations for the detection of cocaine and ATP [11]. Using quantum dot (QD) in place of methylene blue, Yuan et al. developed a "signal on" and sensitive electrochemical sandwich assay for the detection of cocaine and ATP [12]. This assay obtained detection limits of 50 nM and 30 nM for cocaine and ATP, respectively. An electrochemical sandwich assay, in which β-cyclodextrin (β-CD) was modified on the glassy carbon electrode (GCE) for combining with diethylstilbestrol and platinum nanoparticles (PtNPs) coupled with horseradish peroxidase (HRP) were immobilized on the polymerase chelate for signal amplification, was developed to detect diethylstilbestrol with a detection limit of 0.30 pg mL$^{-1}$ [13]. In milk samples, this assay showed a good detection result, which was consistent with that of traditional HPLC method. Bisphenol A, a serious environmental contaminant, could be detected using aptamer sandwich-based carbon nanotube sensor with a detection limit of 10 fM [14].

### 10.2.2   Electrochemiluminescence Sandwich Assays

Electrochemiluminescence (ECL), a kind of transduction technology with high sensitivity and selectivity, has been widely used in the detection of environmental pollutants, pharmaceuticals, biomarkers, macromolecules, small molecules and ions [15]. Xing et al. proposed an ultrasensitive sandwich ECL assay with aptamer-modified ECL nanopores for the detection of ATP, which possessed a detection limit of 10 pM [16]. In their project, tris(2,2'-bipyridyl)ruthenium (TBR)-cysteamine-modified gold nanoparticles (AuNPs) was used as barcodes for signal amplification. ATP bound two affinity aptamers to form the sandwich complex in which the targets were captured by micromagnetic particles on the surface of the electrode and quantified by ECL intensity. A "signal-off" sandwich ECL assay was developed for the detection of ATP with a detection limit of 0.03 pmol L$^{-1}$ [17]. ATP induced the formation of the sandwich complex with ferrocene tag, leading to a quenching of Ru(bpy)$_3^{2+}$. Recently, Chen et al. constructed an ECL sensor through

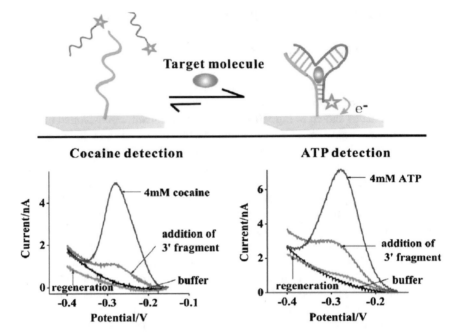

**Fig. 10.1** Schematic illustration of an electrochemical sandwich assay for the detection of small molecules such as cocaine and ATP. Single aptamer sequences are capable of detecting small molecule targets on the gold electrode surface. Target binding will lead to a large increase in faradaic current (Reprinted with the permission from Ref. [10]. Copyright 2009 American Chemical Society)

a dual-molecular recognition and the quenching effect of polyaniline (PANi), as shown in Fig. 10.2 [18]. The gold nanoparticle-functionalized $g$–$C_3N_4$ nanosheets (Au–$g$–$C_3N_4$ NS) as a matrix immobilized dithiobis-(succinimidyl propionate) (DSP, a recognition element) that conjugated with dopamine. 3-Aminophenyl-boronic acid-functionalized PANi (APBA/PANi) was anchored on the surface of GCE via phenylboronic acid-diol specificity to form the sandwich complex. PANi quenched the ECL emission of $g$–$C_3N_4$ to cause the significant decrease of ECL signal. This assay showed a high sensitivity and selectivity, and a detection limit of 0.033 pM for the detection of dopamine was achieved.

## 10.2.3 Fluorescence Sandwich Assays

Fluorescence resonance energy transfer (FRET) is an old physical phenomenon that has become a useful technique for probing intermolecular interactions and determining the spatial extension in biomedical research and drug discovery [19]. Fluorescent biosensors based on FRET have also been developed for biosensing

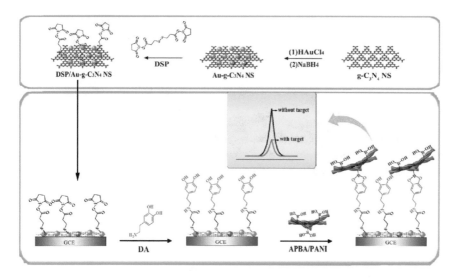

**Fig. 10.2** Schematic illustration of an electrochemiluminescence sandwich assay for the detection of small molecules such as dopamine. Gold nanoparticle-functionalized graphite-like carbon nitride nanosheets (Au–g–$C_3N_4$ NS) with strong electrochemiluminescence intensity as a matrix immobilize recognition element dithiobis-(succinimidyl propionate) (DSP), capturing dopamine (DA). 3-Aminophenylboronic acid-functionalized PANI (APBA/PANI) was incubated onto the electrode to achieve an ECL sensor with a sandwich structure (Reprinted with the permission from Ref. [18]. Copyright 2017 Elsevier)

applications with high sensitivity and good stability [20]. Wang et al. reported a highly selective sandwich FRET assay for the detection of ATP through the high specific recognition of silica-coated upconverting nanoparticles (Si@UCNPs) to ATP aptamer, as shown in Fig. 10.3 [21]. Aptamer I was covalently conjugated to Si@UCNPs. With the addition of ATP, aptamer I and black hole quencher-1 (BHQ1)-labeled aptamer were hybridized to form the sandwich complex on the surface of Si@UCNPs. Under 980 nm laser illumination, energy transfer between Si@UCNPs (the donor) and BHQ1 (the acceptor) was quantitative to the concentration of ATP. This assay detected ATP with a detection limit of 1.70 μM through determining the fluorescence change of Si@UCNPs at 550 nm. A simple sandwich-type FRET assay was developed using BHQ-labeled aptamer as the quencher and 6-FAM-labeled aptamer as the fluorophore. The quencher and fluorophore were bound together on the surface of 19-nortestosterone to detect 19-nortestosterone with a detection limit of 5 μM through FRET quenching in a homogenous solution [22]. Polymerase chain reaction (PCR), a powerful signal amplification technology, is employed in the construction of a sandwich fluorescence assay for the detection of 3-phenoxybenzoic acid with a detection limit of 20 pg mL$^{-1}$ [23] and 17β-estradiol with a detection limit of 16 pg mL$^{-1}$ [24]. Macrocyclic host dye, which was formed from the interaction of dansyl-labeled

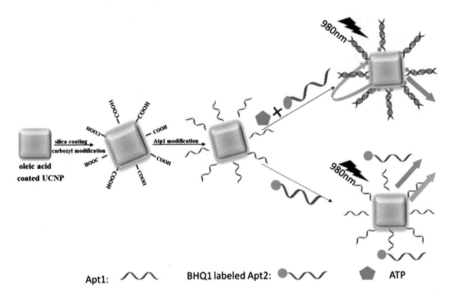

**Fig. 10.3** Schematic illustration of a FRET sandwich assay for the detection of small molecules such as ATP. The Si@UCNPs as the energy donor are covalently modified with one of the aptamer fragments (Apt1). BHQ1-labeled Apt2 is hybridized in the presence of ATP to induce energy transfer (Reprinted with the permission from Ref. [21]. Copyright 2013 Elsevier)

DNA and β-cylodextrin-modified DNA in the presence of adenosine, was proposed to fabricate a sensitive sandwich fluorescence assay for the detection of adenosine with a detection limit of 1 μM [25].

## 10.2.4 Colorimetric Sandwich Assays

Colorimetric assays that determine the concentration of chemical substances in a solution through the color change in the presence of the analyte have been widely used in medical laboratories to test enzymes, antibodies, and many other analytes. They have been also widely adopted for industrial purposes, for example, to analyze water samples in industrial water treatment [26]. AuNP-based colorimetric assays have drawn wide attention in diagnostic applications because of their simplicity and versatility [27]. A sensitive sandwich colorimetric assay in which ATP specifically recognized AuNP-modified aptamer and biotin-labeled aptamer to generate the sandwich complex was developed for the detection of ATP with a detection limit of 0.5 μM [28]. The sandwich structure with ATP specifically recognized AuNPs led to a significant increase in optical signals.

Colorimetric assays based on a traditional enzyme have been widely reported for the detection of small molecules [29–35]. As an example, Ravelet et al. reported a "signal-on" sandwich-type colorimetric assay for the detection of adenosine, as

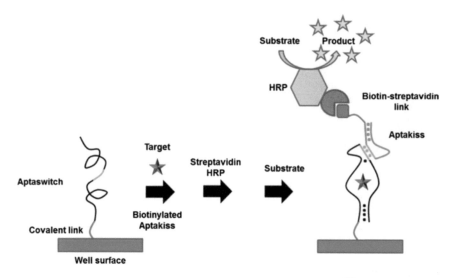

**Fig. 10.4** Schematic illustration of a colorimetric sandwich assay for the detection of small molecules such as adenosine. The aptaswitch is immobilized onto the well surface to capture the target (Reprinted with the permission from Ref. [30]. Copyright 2016 American Chemical Society)

shown in Fig. 10.4 [30]. The aptamer was first immobilized on the surface of a microplate well to capture the target. The folded aptamer bound to the biotinylated aptakiss through loop–loop interaction. The streptavidin-HRP was then conjugated to form the sandwich complex. The catalytic reaction of 3,3′,5,5′-tetra-methylbenzidine by HRP produced a blue-colored solution, which could be observed by naked eyes. Using the similar strategy, malachite green and leuco-malachite green with detection limits of 7.02 and 0.55 ng mL$^{-1}$, respectively, were also detected with a good correlation in comparison with the data of HPLC [29].

## 10.2.5   Other Sandwich Assays

In addition to the classical sandwich assays discussed above, there are also other assays based on the sandwich complex including chemiluminescence assay, surface plasmon resonance (SPR) assay, surface-enhanced Raman scattering (SERS) assay. These assays have been applied to the quantification of small molecules. For example, sandwich chemiluminescence assay based on the antigen-dependent sta-bilization of the antibody variable region was employed for the detection of 11-deoxycortisol with a detection limit of 9 fmol/assay [36]. Sandwich SPR assay using NIR-streptavidin-coated quantum dots as a nano-enhanced SPR imaging sensor was developed for the detection of progesterone with a detection limit of 5 nM in phosphate buffer [37]. Sandwich SERS assay with single functional magnetic bead as universal biosensing platform was demonstrated for the detection

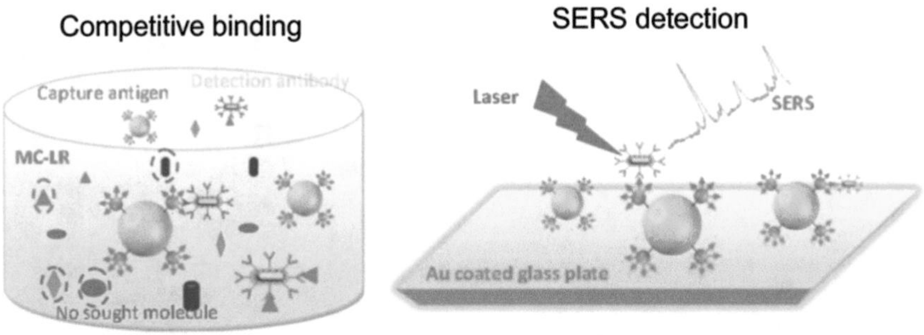

**Fig. 10.5** Schematic illustration of a SERS sandwich assay for the detection of small molecules such as microcystin-leucine-arginine and bisphenol A. The magnetic beads modified with coated antigen bind with SERS nanobioprobes. The complex is dispersed on Au-coated glass plate and the SERS intensity is detected (Reprinted with the permission from Ref. [38]. Copyright 2016 Elsevier)

of microcystin-leucine-arginine and bisphenol A with a detection limit of 0.01 and 0.03 $\mu g\ L^{-1}$, respectively, as shown in Fig. 10.5 [38]. The sandwich complex of Au@Raman reporters@Ag core–shell nanorods (Au@RR@Ag NRs) was formed via layer-by-layer as the detection antibody. The magnetic beads were modified with bovine serum albumin (BSA) as the capture antigen. The analyte bound the detection antibody and the capture antigen together to form the sandwich complex in a solution. The complex was dispersed on Au-coated glass plate, and the SERS detection for small molecules was carried out on this Au-coated substrate.

## 10.3 Sandwich Assays for Ion Detection

### 10.3.1 Electrochemical Sandwich Assays

A sandwich electrochemical assay for the detection of $Hg^{2+}$ was reported through hybridization chain reaction (HCR) of DNA coupled with Ag@Au core–shell nanoparticles (Ag@Au CSNPs) amplification, as shown in Fig. 10.6 [39]. Graphene–nafion film was performed on the surface of GCE, and then AuNPs were deposited to immobilize the capture probe (CP). In the presence of $Hg^{2+}$, CP hybridized with the detection probe (DP) by $T–Hg^{2+}–T$ and further triggered HCR of two-ferrocene (Fc)-labeled hairpin DNA to form dsDNA extension. Ag@Au CSNPs were then adsorbed to amplify the electrochemical signal. This assay had a high selectivity for the detection of $Hg^{2+}$ with a detection limit of 3.6 pM. With the aid of $C–Ag^+–C$ using cytosine-rich DNA oligonucleotide probes, a sandwich electrochemical assay through signal amplification of silver enhancement was developed for the detection of $Ag^+$ with a detection limit of 2 pM [40]. Using crown ethers as high-affinity binding receptors, a sandwich electrochemical assay

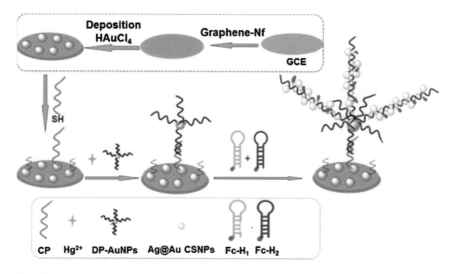

**Fig. 10.6** Schematic illustration of an electrochemical sandwich assay for the detection of ions such as $Hg^{2+}$. AuNPs are deposited on the modified GCE for the immobilization of capture probe. The detection probe-modified AuNPs are hybridized to open two-ferrocene (Fc)-modified hairpin DNA. Ag@Au core–shell nanoparticles (Ag@Au CSNPs) are adsorbed to amplify the electrochemical signal (Reprinted with the permission from Ref. [39]. Copyright 2016 Elsevier)

was constructed via hydrogen bonding and moiety interaction on the surface of gold electrode to detect $Cr^{6+}$ in drinking water with a detection limit of 0.0014 ppb [41]. Furthermore, the sandwich complex constructed in nanopores of porous anodic alumina membrane was used for the detection of $Fe(CN)_6^{3-}$ with a detection limit of 1 pM using AuNPs as electrochemical signal amplification [42].

## 10.3.2 Fluorescence Sandwich Assays

Chen et al. reported a simple sandwich fluorescence assay with the formation of T–$Hg^{2+}$–T using single fluorophore-labeled poly(dT) as the donor for the detection of $Hg^{2+}$ with a detection limit of 20 nM [43]. The sandwich complex of T–$Hg^{2+}$–T was absorbed on the surface of graphene oxide sheets that led to the fluorescence quenching; therefore, $Hg^{2+}$ was detected with a good selectivity showing a detection limit of 0.92 nM [44]. T–$Hg^{2+}$–T triggered a sandwich-type system of UCNPs as the donor and AuNRs as the acceptor. The sandwich assay could detect $Hg^{2+}$ with a detection limit of 2 nM, as shown in Fig. 10.7 [45]. UCNPs had a near-infrared excitation wavelength of 980 nm and a near-infrared emission wavelength of 804 nm. AuNRs had good absorption ability for around 806 nm band. The sandwich complex caused significant decrease of fluorescence intensity

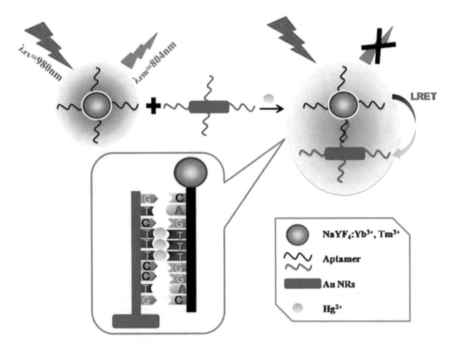

**Fig. 10.7** Schematic illustration of a fluorescence sandwich assay for the detection of ions such as $Hg^{2+}$. The upconverting $NaYF_4:Yb^{3+}$, $Tm^{3+}$ nanoparticles and gold nanorods with the $Hg^{2+}$ aptamer are conjugated together due to the formation of $T-Hg^{2+}-T$ in the presence of $Hg^{2+}$ (Reprinted with the permission from Ref. [45]. Copyright 2013 The Royal Society of Chemistry)

with the addition of $Hg^{2+}$ due to luminescence resonant energy transfer (LRET). FRET technique was also used in 1,8-naphthalimide-Rhodamine-based sensors for the detection of $Hg^{2+}$ [46].

## 10.3.3  Colorimetric Sandwich Assays

Because of the specific optical properties, colloidal solution of AuNPs has obvious color change in the visible region when AuNPs are aggregated [47]. To date, AuNPs have played a critical role in the development of biosensors, particularly in colorimetric assays for the detection of metal ions [48]. Azacrown ether-modified AuNPs could selectively capture $Pb^{2+}$ to form the sandwich coordination, resulting in the AuNPs aggregation, and then obvious color change from brown to purple was observed. This mechanism was employed in a simple and fast colorimetric assay for the detection of $Pb^{2+}$ at room temperature [49]. Pyridine-modified AuNPs selectively capturing $Cu^{2+}$ or $Ag^+$ yielded the nanoparticle aggregation, and then a

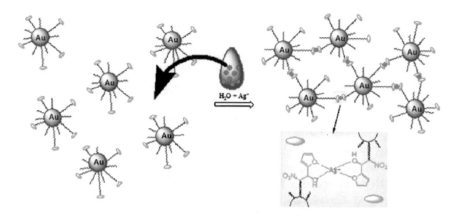

**Fig. 10.8** Schematic illustration of a colorimetric sandwich assay for the detection of ions such as Ag$^+$. The recognition mechanism is attributed to the formation of a sandwich between Ag(I) ion and two furfuryl alcohol moieties that are attached to separate nanoparticles (Reprinted with the permission from Ref. [51]. Copyright 2016 Springer-Verlag Wien)

color change was monitored by the naked eye [50]. A colorimetric assay for the detection of Ag$^+$ was described on the basis of the formation of the sandwich structure between Ag$^+$ and two furfuryl alcohol groups that are modified on the surface of AuNPs, as shown in Fig. 10.8 [51]. The detectable color change from pale brown to dark blue was resulted from the nanoparticle aggregation. This simple and fast assay had a detection limit of 12 nM.

Furthermore, T–Hg$^{2+}$–T coordination of DNA-immobilized AuNPs and DNA-immobilized magnetic microparticles was bound to form the sandwich-type structure, which was demonstrated for the detection of Hg$^{2+}$ in urine and drinking water with a detection limit of 0.09 μg mL$^{-1}$ [52]. T–Hg$^{2+}$–T coordination of multi-walled carbon nanotubes (MWCNTs) with the optical properties on the lateral flow strip was successfully constructed to a strip biosensor for the visible detection of Hg$^{2+}$ with a detection limit of 0.05 ppb without instrumentation [53]. Notably, this biosensor worked without any instrumentation. Polydiacetylenes (PDAs)-capped semiconducting polymer dots (Pdots) modified by crown ether specifically recognized Pb$^{2+}$ by the formation of crown-Pb$^{2+}$-carboxylate sandwich complex on the surface of Pdots [54]. The perturbed and strained PDA caused a chromatic change of PDA from blue to red observed by naked eyes. Hg$^{2+}$ inserted the couple of 1,8-naphthalimide-Rhodamine along with an apparent color change from yellow to orange/pink. Accordingly, a colorimetric sensor for the detection of Hg$^{2+}$ in corresponding solvent systems was fabricated [46].

### 10.3.4  Other Sandwich Assays

Other assays that were applied for the detection of ions through the sandwich complex contained quartz crystal microbalance (QCM) assay, SERS assay, etc. The QCM technique based on the piezoelectric properties of quartz crystals is a powerful tool for the quantification of molecular interaction [55]. QCM-based sensors have been widely used in a series of application for the detection of biomarkers, macromolecules, small molecules and ions [56]. The DNA hybridization based on T–Hg$^{2+}$–T on the surface of QCM led to the frequency change that was corresponding to the mass change in the QCM crystal, which was developed for the detection of Hg$^{2+}$ [57–59]. A SERS biosensor was developed by hybridizing gold nanostar@Raman-reporter@silica sandwich complex-modified DNA as the SERS probe with DNA immobilized on the surface of gold nanohole array for the detection of Hg$^{2+}$ or Ag$^+$ in human saliva, as shown in Fig. 10.9 [60]. The formation of T–Hg$^{2+}$–T or C–Ag$^+$–C triggered the DNA hybridization that caused the nanoparticle aggregation on the surface of gold nanohole array. The electromagnetic field was enhanced, thus amplifying the SERS signal. Recently, Au bowtie nanoarrays/n-layer graphene/AuNPs sandwich structure was fabricated and applied to detect Hg$^{2+}$ in water by T–Hg$^{2+}$–T coordination with a detection limit of 8.3 nM [61].

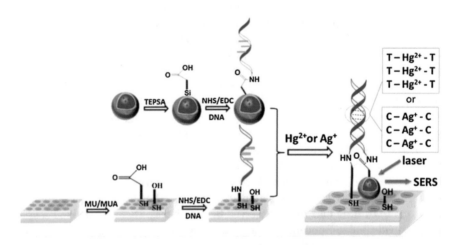

**Fig. 10.9**  Schematic illustration of a SERS sandwich assay for the detection of ions such as Hg$^{2+}$ or Ag$^+$. The gold nanostar@Raman-reporter@silica sandwich nanoparticles and Au nanohole arrays are functionalized with ssDNA sequences. The two ssDNA sequences hybridize only when Hg$^{2+}$ or Ag$^+$ is present due to T–T or C–C mismatch (Reprinted with the permission from Ref. [60]. Copyright 2015 The Royal Society of Chemistry)

## 10.4   Conclusion

In this chapter, we have highlighted the recent development of some sandwich assays for the detection of small molecules and ions. Several techniques such as electrochemical method, electrochemiluminescence method, fluorescence method, colorimetric method, surface plasmon resonance (SPR), surface-enhanced Raman scattering (SERS), and quartz crystal microbalance (QCM) have been employed in the sandwich assays. It is worth noting that every technique has their advantages and disadvantages in the fabrication and practical application of the sandwich assays. For example, sandwich colorimetric assays have a simple fabrication process but the sensitivity of sandwich colorimetric assays is relatively low in comparison with other techniques. The sensitivity and specificity of the present sandwich assays in detecting small molecules and ions are satisfactory and acceptable. However, almost all sandwich assays still stay in the bench of laboratory and there is a long way to apply the sandwich assays into the market. With continued efforts, it is strongly believed that the product will be sold in the market in the near future.

## References

1. Wang XH, Wang S (2008) Sensors and biosensors for the determination of small molecule biological toxins. Sensors 8:6045–6054
2. Wang HM, Feng ZQQ, Xu B (2017) Bioinspired assembly of small molecules in cell milieu. Chem Soc Rev 46:2421–2436
3. Liu DB, Wang Z, Jiang XY (2011) Gold nanoparticles for the colorimetric and fluorescent detection of ions and small organic molecules. Nanoscale 3:1421–1433
4. Feng CJ, Dai S, Wang L (2014) Optical aptasensors for quantitative detection of small biomolecules: a review. Biosens Bioelectron 59:64–74
5. Qian XH, Xu ZC (2015) Fluorescence imaging of metal ions implicated in diseases. Chem Soc Rev 44:4487–4493
6. Zhang JF, Zhou Y, Yoon J, Kim JS (2011) Recent progress in fluorescent and colorimetric chemosensors for detection of precious metal ions (silver, gold and platinum ions). Chem Soc Rev 40:3416–3429
7. Duong TQ, Kim JS (2010) Fluoro- and chromogenic chemodosimeters for heavy metal ion detection in solution and biospecimens. Chem Rev 110:6280–6301
8. Liu Y, Deng Y, Dong HM, Liu KK, He NY (2017) Progress on sensors based on nanomaterials for rapid detection of heavy metal ions. Sci China-Chem 60:329–337
9. Gumpu MB, Sethuraman S, Krishnan UM, Rayappan JBB (2015) A review on detection of heavy metal ions in water—an electrochemical approach. Sens Actuator B-Chem 213:515–533
10. Zuo XL, Xiao Y, Plaxco KW (2009) High specificity, electrochemical sandwich assays based on single aptamer sequences and suitable for the direct detection of small-molecule targets in blood and other complex matrices. J Am Chem Soc 131:6944–6945
11. Zhao T, Liu R, Ding XF, Zhao JC, Yu HX, Wang L, Xu Q, Wang X, Lou XH, He M, Xiao Y (2015) Nanoprobe-enhanced, split aptamer-based electrochemical sandwich assay for ultrasensitive detection of small molecules. Anal Chem 87:7712–7719

12. Zhang HX, Jiang BY, Xiang Y, Zhang YY, Chai YQ, Yuan R (2011) Aptamer/quantum dot-based simultaneous electrochemical detection of multiple small molecules. Anal Chim Acta 688:99–103

13. Yan ZD, Xiong P, Gan N, He JL, Long NB, Cao YT, Hu FT, Li TH (2015) A novel sandwich-type noncompetitive immunoassay of diethylstilbestrol using beta-cyclodextrin modified electrode and polymer-enzyme labels. J Electroanal Chem 736:30–37

14. Lee J, Jo M, Kim TH, Ahn JY, Lee DK, Kim S, Hong S (2011) Aptamer sandwich-based carbon nanotube sensors for single-carbon-atomic-resolution detection of non-polar small molecular species. Lab Chip 11:52–56

15. Li LL, Chen Y, Zhu JJ (2017) Recent advances in electrochemiluminescence analysis. Anal Chem 89:358–371

16. Zhou XM, Duan RX, Xing D (2012) Highly sensitive detection of protein and small molecules based on aptamer-modified electrochemiluminescence nanoprobe. Analyst 137:1963–1969

17. Li M, Yang HM, Ma C, Zhang Y, Ge SG, Yu JH, Yan M (2014) A sensitive signal-off aptasensor for adenosine triphosphate based on the quenching of $Ru(bpy)_3^{2+}$-doped silica nanoparticles electrochemiluminescence by ferrocene. Sens Actuator B-Chem 191:377–383

18. Zuo FM, Jin L, Fu XM, Zhang H, Yuan R, Chen SH (2017) An electrochemiluminescent sensor for dopamine detection based on a dual-molecule recognition strategy and polyaniline quenching. Sens Actuator B-Chem 244:282–289

19. Sapsford KE, Berti L, Medintz IL (2004) Fluorescence resonance energy transfer—concepts, applications and advances. Minerva Biotechnol 16:247–273

20. Chen GW, Song FL, Xiong XQ, Peng XJ (2013) Fluorescent nanosensors based on fluorescence resonance energy transfer (FRET). Ind Eng Chem Res 52:11228–11245

21. He XX, Li ZX, Jia XK, Wang KM, Yin JJ (2013) A highly selective sandwich-type FRET assay for ATP detection based on silica coated photon upconverting nanoparticles and split aptamer. Talanta 111:105–110

22. Bai WH, Zhu C, Liu JC, Yan MM, Yang SM, Chen AL (2016) Split aptamer-based sandwich fluorescence resonance energy transfer assay for 19-nortestosterone. Microchim Acta 183:2533–2538

23. Kim HJ, McCoy M, Gee SJ, Gonzalez-Sapienza GG, Hammock BD (2011) Noncompetitive phage anti-immunocomplex real-time polymerase chain reaction for sensitive detection of small molecules. Anal Chem 83:246–253

24. Dong JH, Hasan S, Fujioka Y, Ueda H (2012) Detection of small molecule diagnostic markers with phage-based open-sandwich immuno-PCR. J Immunol Methods 377:1–7

25. Yang C, Spinelli N, Perrier S, Defrancq E, Peyrin E (2015) Macrocyclic host-dye reporter for sensitive sandwich-type fluorescent aptamer sensor. Anal Chem 87:3139–3143

26. Chen WW, Guo YM, Zheng WS, Xianyu YL, Wang Z, Jiang XY (2014) Recent progress of colorimetric assays based on gold nanoparticles for biomolecules. Chin J Anal Chem 42:307–314

27. Zhao W, Brook MA, Li YF (2008) Design of gold nanoparticle-based colorimetric biosensing assays. ChemBioChem 9:2363–2371

28. Zhu C, Zhao Y, Yan MM, Huang YF, Yan J, Bai WH, Chen AL (2016) A sandwich dipstick assay for ATP detection based on split aptamer fragments. Anal Bioanal Chem 408:4151–4158

29. Dong JX, Xu C, Wang H, Xiao ZL, Gee SJ, Li ZF, Wang F, Wu WJ, Shen YD, Yang JY, Sun YM, Hammock BD (2014) Enhanced sensitive immunoassay: noncompetitive phage anti-immune complex assay for the determination of malachite green and leucomalachite green. J Agric Food Chem 62:8752–8758

30. Chovelon B, Durand G, Dausse E, Toulme JJ, Faure P, Peyrin E, Ravelet C (2016) ELAKCA: Enzyme-linked aptamer kissing complex assay as a small molecule sensing platform. Anal Chem 88:2570–2575

31. Hara Y, Dong J, Ueda H (2013) Open-sandwich immunoassay for sensitive and broad-range detection of a shellfish toxin gonyautoxin. Anal Chim Acta 793:107–113

32. Kubota K, Mizukoshi T, Miyano H (2013) A new approach for quantitative analysis of L-phenylalanine using a novel semi-sandwich immunometric assay. Anal Bioanal Chem 405:8093–8103
33. Sharma AK, Kent AD, Heemstra JM (2012) Enzyme-linked small-molecule detection using split aptamer ligation. Anal Chem 84:6104–6109
34. Quinton J, Charruault L, Nevers MC, Volland H, Dognon JP, Creminon C, Taran F (2010) Toward the limits of sandwich immunoassay of very low molecular weight molecules. Anal Chem 82:2536–2540
35. Kobayashi N, Oyama H, Suzuki I, Kato Y, Umemura T, Goto J (2010) Oligosaccharide-assisted direct immunosensing of small molecules. Anal Chem 82:4333–4336
36. Ihara M, Suzuki T, Kobayashi N, Goto J, Ueda H (2009) Open-sandwich enzyme immunoassay for one-step noncompetitive detection of corticosteroid 11-deoxycortisol. Anal Chem 81:8298–8304
37. Zeidan E, Shivaji R, Henrich VC, Sandros MG (2016) Nano-SPRi aptasensor for the detection of progesterone in buffer. Sci Rep 6:26714
38. Xiao R, Wang CW, Zhu AN, Long F (2016) Single functional magnetic-bead as universal biosensing platform for trace analyte detection using SERS-nanobioprobe. Biosens Bioelectron 79:661–668
39. Li ZB, Miao XM, Xing K, Peng X, Zhu AH, Ling LS (2016) Ultrasensitive electrochemical sensor for $Hg^{2+}$ by using hybridization chain reaction coupled with Ag@Au core-shell nanoparticles. Biosens Bioelectron 80:339–343
40. Zhang YL, Li HY, Xie JL, Chen M, Zhang DD, Pang PF, Wang HB, Wu Z, Yang WR (2017) Electrochemical biosensor for silver ions based on amplification of DNA-Au bio-bar codes and silver enhancement. J Electroanal Chem 785:117–124
41. Wei J, Guo Z, Chen X, Han DD, Wang XK, Huang XJ (2015) Ultrasensitive and ultraselective impedimetric detection of Cr(VI) using crown ethers as high-affinity targeting receptors. Anal Chem 87:1991–1998
42. Li SJ, Xia N, Yuan BQ, Du WM, Sun ZF, Zhou BB (2015) A novel DNA sensor using a sandwich format by electrochemical measurement of marker ion fluxes across nanoporous alumina membrane. Electrochim Acta 159:234–241
43. Guo LQ, Hu H, Sun RQ, Chen GA (2009) Highly sensitive fluorescent sensor for mercury ion based on photoinduced charge transfer between fluorophore and pi-stacked T–Hg(II)–T base pairs. Talanta 79:775–779
44. Li M, Zhou XJ, Ding WQ, Guo SW, Wu NQ (2013) Fluorescent aptamer-functionalized graphene oxide biosensor for label-free detection of mercury(II). Biosens Bioelectron 41:889–893
45. Chen HQ, Yuan F, Wang SZ, Xu J, Zhang YY, Wang L (2013) Near-infrared to near-infrared upconverting $NaYF4:Yb^{3+}$, $Tm^{3+}$ nanoparticles-aptamer-Au nanorods light resonance energy transfer system for the detection of mercuric(II) ions in solution. Analyst 138:2392–2397
46. Fang Y, Zhou Y, Li JY, Rui QQ, Yao C (2015) Naphthalimide-Rhodamine based chemosensors for colorimetric and fluorescent sensing $Hg^{2+}$ through different signaling mechanisms in corresponding solvent systems. Sens Actuator B-Chem 215:350–359
47. Vilela D, Gonzalez MC, Escarpa A (2012) Sensing colorimetric approaches based on gold and silver nanoparticles aggregation: chemical creativity behind the assay. A review. Anal Chim Acta 751:24–43
48. Priyadarshini E, Pradhan N (2017) Gold nanoparticles as efficient sensors in colorimetric detection of toxic metal ions: a review. Sens Actuator B-Chem 238:888–902
49. Alizadeh A, Khodaei MM, Karami C, Workentin MS, Shamsipur M, Sadeghi M (2010) Rapid and selective lead (II) colorimetric sensor based on azacrown ether-functionalized gold nanoparticles. Nanotechnology 21:315503
50. Alizadeh A, Khodaei MM, Hamidi Z, Bin Shamsuddin M (2014) Naked-eye colorimetric detection of $Cu^{2+}$ and $Ag^+$ ions based on close-packed aggregation of pyridines-functionalized gold nanoparticles. Sens Actuator B-Chem 190:782–791

51. Alizadeh A, Abdi G, Khodaei MM (2016) Colorimetric and visual detection of silver(I) using gold nanoparticles modified with furfuryl alcohol. Microchim Acta 183:1995–2003

52. Hsu IH, Hsu TC, Sun YC (2011) Gold-nanoparticle-based graphite furnace atomic absorption spectrometry amplification and magnetic separation method for sensitive detection of mercuric ions. Biosens Bioelectron 26:4605–4609

53. Yao L, Teng J, Zhu MY, Zheng L, Zhong YH, Liu GD, Xue F, Chen W (2016) MWCNTs based high sensitive lateral flow strip biosensor for rapid determination of aqueous mercury ions. Biosens Bioelectron 85:331–336

54. Kuo SY, Li HH, Wu PJ, Chen CP, Huang YC, Chan YH (2015) Dual colorimetric and fluorescent sensor based on semiconducting polymer dots for ratiometric detection of lead ions in living cells. Anal Chem 87:4765–4771

55. Qiao XX, Zhang XJ, Tian Y, Meng YG (2016) Progresses on the theory and application of quartz crystal microbalance. Appl Phys Rev 3:206–222

56. Vashist SK, Vashist P (2011) Recent advances in quartz crystal microbalance-based sensors. J Sens 571405

57. Sheng ZH, Han JH, Zhang JP, Zhao H, Jiang L (2011) Method for detection of $Hg^{2+}$ based on the specific thymine-$Hg^{2+}$-thymine interaction in the DNA hybridization on the surface of quartz crystal microbalance. Colloid Surf B-Biointerfaces 87:289–292

58. Chen Q, Wu XJ, Wang DZ, Tang W, Li N, Liu F (2011) Oligonucleotide-functionalized gold nanoparticles-enhanced QCM-D sensor for mercury(II) ions with high sensitivity and tunable dynamic range. Analyst 136:2572–2577

59. Dong ZM, Zhao GC (2012) Quartz crystal microbalance aptasensor for sensitive detection of mercury(II) based on signal amplification with gold nanoparticles. Sensors 12:7080–7094

60. Zheng P, Li M, Jurevic R, Cushing SK, Liu YX, Wu NQ (2015) A gold nanohole array based surface-enhanced Raman scattering biosensor for detection of silver(I) and mercury(II) in human saliva. Nanoscale 7:11005–11012

61. Zhang XG, Dai ZG, Si SY, Zhang XL, Wu W, Deng HB, Wang FB, Xiao XH, Jiang CZ (2017) Ultrasensitive SERS substrate integrated with uniform subnanometer scale "hot spots" created by a graphene spacer for the detection of mercury ions. Small 13:1603347

# Chapter 11
# Sandwich Assay for Pathogen and Cells Detection

## Shaoguang Li, Hui Li and Fan Xia

**Abstract** Sandwich assay biosensors make it possible to detect bacterial pathogens and cancer cells at extremely low level. In this chapter, we have summarized the recent developments of sandwich assay for pathogen and whole-cell detection using a variety of techniques. In particular, we highlighted some of the most common techniques in sandwich assay biosensors such as optics-based detection, electrochemistry-based detection, and mechanics-based detection.

**Keywords** Whole-cell · Pathogen · Bacterial · Signal amplification
Sandwich assay

The original version of this chapter was revised: Foreword has been included and authors' affiliations have been updated. The erratum to this chapter is available at https://doi.org/10.1007/978-981-10-7835-4_13

S. Li (✉) · H. Li · F. Xia
Engineering Research Center of Nano-Geomaterials of Ministry of Education,
Faculty of Materials Science and Chemistry, China University of Geosciences,
Wuhan 430074, People's Republic of China
e-mail: lishaoguang@cug.edu.cn

H. Li
e-mail: lihui-chem@cug.edu.cn

F. Xia
e-mail: xiafan@cug.edu.cn; xiafan@hust.edu.cn

F. Xia
Hubei Key Laboratory of Bioinorganic Chemistry & Materia Medica,
School of Chemistry and Chemical Engineering, Huazhong University of Science
and Technology, Wuhan 430074, People's Republic of China

© Springer Nature Singapore Pte Ltd. 2018
F. Xia et al. (eds.), *Biosensors Based on Sandwich Assays*,
https://doi.org/10.1007/978-981-10-7835-4_11

## 11.1 Introduction

In the past several years, there are significant developments of diagnostic techniques for public health, food and water safety, and homeland security [1, 2]. In particular, plenty of methods and techniques have vastly advanced the detection of pathogens, cancer cells, and other disorders, namely phenotypic, immunological, molecular, and genotypic protocols [3–5]. Nevertheless, many of these techniques are conventional, laboratory-based diagnostic methods, which require long processing time, specialized equipment and are tedious to perform. As such, the demand for sensitive, selective, rapid, and cost-effective detection of bacterial pathogens and cancer cells is highly increasing [6–8].

The sandwich assay biosensors can fill this role because they are highly specific and reproducible to a variety of biological structures, organisms, and processes [9–11]. Moreover, easy signal amplification of sandwich assay promotes them to be with great sensitivity compared with other platforms. As such, this assay has been extensively applied to a variety of analytes, such as metal ions, small molecules, nucleic acids, proteins, and bacterial pathogens and cells [12, 13].

In terms of pathogen and cancer cell sensing, one of the well-established strategies is the detection of their biomolecule components. These components, including DNA [14, 15], RNA [16, 17], proteins [18], and exotoxins [19], have been successfully detected at exceedingly low levels by sandwich assay using polymerase chain reaction (PCR) or immunoassays techniques. The major disadvantage of this component-detecting strategy is the requirement for sample pre-enrichment, sample processing, expensive instruments, and commercial reagents. To solve this issue, sandwich assay biosensors for whole-cell detection, again, without any sample processing, are much more desirable for accurate, rapid, and cost-effective testing especially for the point-of-care detection. Additionally, whole-cell detection approach also provides the possibility of real-time monitoring of the activities of living pathogens and cancerous cells, which helps in elucidation of their functions in a developmental manner [20].

Significant efforts have been devoted in the development of whole-cell detection based on sandwich assay. In a typical sandwich assay, such as the enzyme-linked immunosorbent assay (ELISA) [21], two antibodies bind to one single target at two distinct sites to form a sandwich complex, which leads to highly specific recognition. Upon the sandwich formation, depending on the enzyme catalytic or amplified signaling mechanism, a measurable change in signals is produced and thus the target whole cell can be detected. The utilization of molecular recognition agents such as antibody, aptamer, polypeptide, and bacteriophage has been employed successfully for specifical detection of whole cells [22]. Likewise, some small molecular compounds, such as antibiotics and carbohydrates have been employed as recognition receptor for whole cells [23, 24]. The signaling mechanism has also been extensively expanded along with the development of nanomaterials. In recent years, many promising techniques have been developed and applied to nondestructive whole-cell sensing, such as optical techniques [including colorimetric analysis, fluorescence, surface plasma resonance (SPR)], electro-chemical and mechanical techniques. As the sandwich assays for whole-cell sensing is vast and new works generate constantly, here, we intend to summarize comprehensively the latest advances of this field in general, in support to spur additional ideas in this area.


## 11.2  Optical Detection

As one of the most popular protocols, the optical whole-cell biosensor combining the nondestructive recognition event with optical measurements is of particular interests due to the highly specific bonding, profound signal amplification, visible radiation, and low detection limit. As such, it has been developed vastly based on a variety of spectroscopic techniques. Herein we discuss the colorimetric analysis, fluorescence, and SPR, which are most commonly used for whole-cell detections.


### 11.2.1  Colorimetric Analysis

The colorimetric analysis has attracted a lot of interest due to its visible radiation, low cost, quick feedback, and the possibility of avoiding any expensive instrument. Current studies on pathogens, cancer cells sensing by colorimetric methods aim for achieving a more specific, easy to use, more portable, and low-cost analytical system.

Toward this goal, many research works focused on sandwich assay-based biosensor coupled with nanomaterials for signal amplification [25]. For example, Zhang et al. have developed a nanoparticle cluster (NPC)-based amplification biosensor for the detection of *Listeria monocytogenes*, which is a highly pathogenic foodborne bacterial (Fig. 11.1) [26]. Specifically, they used a glycopeptide antibiotic, vancomycin (Van) as the first recognition agent to capture the cell wall of the pathogen. The aptamer-labeled $Fe_3O_4$ NPC was used as the signal amplification probe, which was also recognized to the pathogen. The sandwich recognition showed high specificity, in which the NPC-based method displayed higher sensitivity than the NP-based method due to its improved catalytic activity [27, 28]. Using this new method, the *L. monocytogenes* cells could be detected within a linear range of $5.4 \times 10^3$ to $10^8$ CFU/mL and a visual limit of detection (LOD) of $5.4 \times 10^3$ CFU/mL [26]. Likewise, Jain et al. have recently demonstrated a surface aminated polycarbonate membrane (PC)-enhanced sandwich assay for *Salmonella typhi* detection. A detection limit of $2 \times 10^3$ cells/ml of bacteria has been achieved with high immobilization efficiency [29].

Gold nanoparticles (AuNPs) have been applied as color developing moiety in numerous colorimetric bioassays [30, 31]. The aggregation of AuNPs usually lead to a distinct color change from red to blue and thus promise for target detection

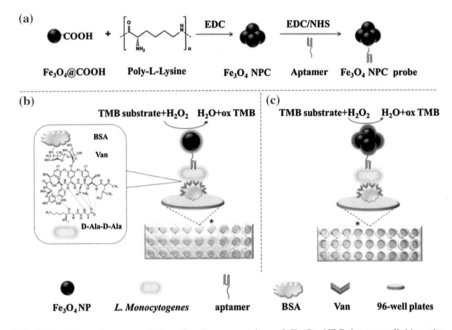

Fig. 11.1 Schematic representation for the preparation of $Fe_3O_4$ NPC by cross-linking the individual mother nanoparticle with poly-L-lysine (**a**), the principle of the $Fe_3O_4$ NP-based biosensor (**b**), and the $Fe_3O_4$ NPC-catalyzed signal amplification biosensor (**c**) (Reprinted with the permission from Ref. [26]. Copyright 2016 Elsevier)

including whole cells [32]. Lu and coworkers have developed a modified AuNPs nanoprobe for colorimetric signal amplification in the detection of *Salmonella enterica*. The optimized LOD is $10^3$ CFU/mL, and their technique has been demonstrated the success of target detection in milk samples with high degree of accuracy (>90%) [33]. Xiong and coworkers have recently established an improved sandwich plasmonic ELISA (pELISA) for determination of *L. monocytogenes* by combining the sandwich ELISA technique with a novel signal-generation mechanism, the catalase (CAT)-mediated growth of plasmonic AuNPs [34], exhibiting an ultralow LOD value at $8 \times 10^0$ CFU/mL (Fig. 11.2) [35].

## 11.2.2  Fluorescence

Fluorescence detection, in contrast to colorimetric assay, is particularly attractive for bacterial pathogens and cancer cells sensing, due to their high-to-signal ratio and improved sensitivity. The commonly used signal transducers are organic dyes (see Fig. 11.3) [36–38] and fluorescent nanoparticles [39].

One of the objective in this area is to develop high-specific, easily implementable bioassay that can be applied to detection and identification of whole

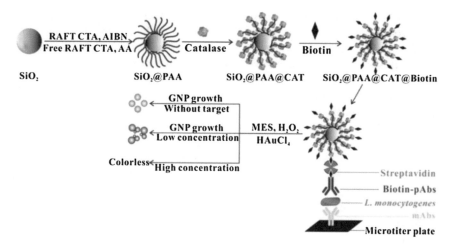

**Fig. 11.2** Schematic of the proposed quantitative immunoassay based on SiO₂@PAA @CAT-catalyzed growth of AuNPs. Specifically, the synthetic SiO₂@PAA@CAT complexes coupled with the biotin–streptavidin system were used to construct a sandwich assay for naked-eye determination of *L. monocytogenes* (Reprinted with the permission from Ref. [35]. Copyright 2015 American Chemical Society)

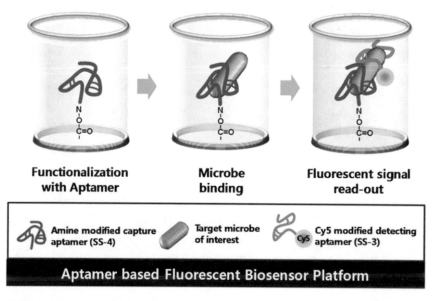

**Fig. 11.3** Principle of *S. sonnei* detection using an aptamer-based fluorescent sandwich-type biosensor platform [Reprinted with the permission from Ref. [38]. Copyright 2017 Multidisciplinary Digital Publishing Institute (MDPI)]

pathogens and cancer cells in complex matrices. Li and coworkers have recently demonstrated a technique for quantitative detection of the *Escherichia coli* O157: H7 (*E. coli* O157:H7) in complex media, which is one of the highly pathogenic agents. Hollow silica nanospheres loading with fluorescein (FHSNs) have been applied to the signal amplification in the sandwich-type immunoassays. Under optimized conditions, this platform provided a sensitive detection of *E. coli* O157: H7 cells with a linear range of 4 to $4 \times 10^8$ CFU/mL and a LOD of 3 CFU/mL. Likewise, this architecture has shown high robustness and high sensitivity for whole-cell sensing in complex sample matrices, such as milk, orange juice, and river water [40]. In another study, Dogan et al. have developed a chitosan-coated CdTe quantum dots (CdTe QDs) as the fluorescence label in the sandwich immunoassays for *E. coli* detection. They achieved a sensitive detection of target in urine matrix and high selectivity over the other four pathogens [41].

Fu and coworkers have recently developed an antibiotic-affinity strategy for fluorimetric detection of *Staphylococcus aureus* (*S. aureus*) cell (Fig. 11.4) [42]. Specifically, the targeted cell was sandwiched by vancocin-modified BSA and fluorescein isothiocyanate (FITC)-labeled antibody. They observed a linear detection from $1.0 \times 10^3$ to $1.0 \times 10^9$ CFU/mL with a LOD of $2.9 \times 10^2$ CFU/mL. Their method exhibited 85–130% of recoveries when applied in spiked apple juice for *S. aureus* detection.

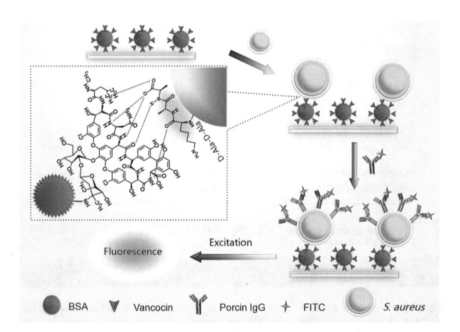

**Fig. 11.4** Principle of sandwich fluorimetric detection of *S. aureus* based on antibiotic-affinity strategy. The target pathogen was captured by vancocin through five-point hydrogen bonds and was further sandwiched by the fluorescein labeled IgG (Reprinted with the permission from Ref. [42]. Copyright 2015 American Chemical Society)

## 11.2.3  Surface Plasmon Resonance

During the past two decades, surface plasma resonance (SPR) techniques have been extensively explored for biosensor platforms targeting pathogens and cells detections, because they are sensitive, label-free and particularly enable the real-time detections of biological targets [43].

Pathogen diagnostics using SPR techniques typically involve signal amplification in order to improve the sensitivity. For example, Eum et al. have developed a SPR-sensing platform for *E. coli* O157:H7 detection. In this study, they immobilized the antibodies onto gold nanorods (GNRs) to enhance the sensitivity of the biosensor. The SPR response with the GNRs labeled antibody was around fourfold improvement of the response than that of from the unlabeled antibody [45]. In another study, Santos et al. have demonstrated the use of SPR to monitor the antibody immobilization protocol for *E. coli* O157:H7 detection [46]. Recently, Liu et al. proposed a SPR immunosensor coupled with antibody-functionalized magnetic nanoparticles (MNPs) for *Salmonella enteritidis* detection (see Fig. 11.5) [44]. Specifically, they immobilized capture antibody via EDC/NHS chemistry onto Au chips and anchored the secondary antibody onto $Fe_3O_4$ MNPs using the same chemistry. This antibody-functionalized MNPs allowed the selective recognition and separation of *S. enteritidis* from the sample matrix under an external magnetic field. This MNPs-enhanced sandwich assay exhibited a large improvement in sensitivity as well as the detection range. Charlermroj et al. compared the sensor performance of a direct, sandwich, or subtractive immunoassay for the detection of

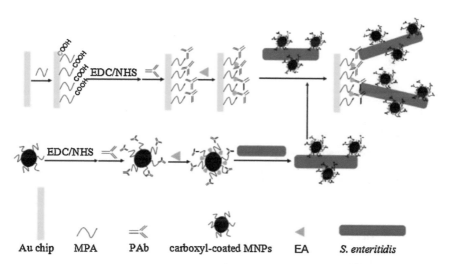

**Fig. 11.5** Schematic representation for the detection of *S. enteritidis* by MNPs-enhanced SPR sandwich assay. The antibody-functionalized MNP acts as both the enrichment reagent of the target and the amplification reagent of SPR immunosensor (Reprinted with the permission from Ref. [44]. Copyright 2016 Elsevier)

bacteria *Acidovorax avenae* subsp. *citrulli* (Aac) and discovered that the direct
assay format exhibited the best sensitivity, while, the sandwich assay provided the
best signal enhancement [47].

As it is commonly seen for SPR-based pathogen detections, nanoparticle amplification is widely employed for cancer cell detections using SPR technique. For
example, Chen et al. reported a sensitive SPR biosensor coupled with MNPs for the
determination of breast cancer cell MCF-7 [48]. The target cancer cells were firstly
captured by the aptamer on the surface, followed by the binding event of
antibody-labeled MNP to form a sandwich assay. As such, the SPR signal enhanced
significantly by MNP immobilization due to the large mass effect and high refractive
index of the assays. With such signal enhancement, this platform exhibited a detection
limit of 500 cells/mL. In a more recent study, Mousavi et al. have developed a
microfluidic chip combined with gold nanoslit SPR for cancer cells detections in
human blood [49]. They coupled this platform with magnetic nanoparticles in support
for efficient immunomagnetic capturing and separation. At last, a LOD of 13 cells/mL
and real-time monitoring of the whole process were achieved (Fig. 11.6).

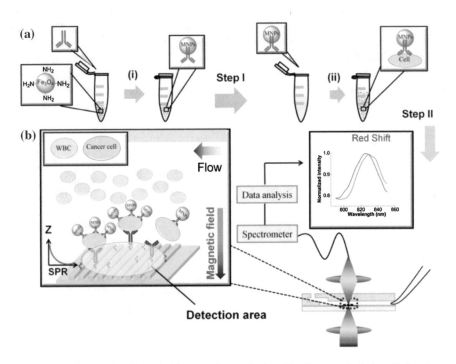

**Fig. 11.6** A schematic of the double capturing method. **a** The first step includes: (i) functionalizing the MNPs with antibody I; (ii) mixing the functionalized MNPs (carrying antibody I) with
the sample to capture the target cells. **b** The second step includes introducing the mixture of blood
sample and MNPs to the microfluidic chip and capturing the MNPs-cells to bind to the antibody II
on the gold nanoslits. The cell binding on the gold nanoslits was monitored by the wavelength shift
of the SPR spectrum [Reprinted with the permission from Ref. [49]. Copyright 2015
Multidisciplinary Digital Publishing Institute (MDPI)]

## 11.3  Electrochemical Detection

The signaling mechanism of electrochemical sandwich assays is based on the electronic communication between the transducer and biomolecules. Because of this unique signaling mechanism, the electrochemical sandwich assays are sensitive, selective, rapid, miniaturizable, and cost-effective, which make them to be of particular interests. They are, for most of cases, more practical for the development of point-of-care devices, especially for the pathogen and cell detections [50].

Electrochemical sandwich-type biosensors for whole-cell detections are typically composed of three components: capture element, target cells, and signal transducer elements. Capture elements are usually DNA/RNA aptamers or antibodies, which are used for anchoring the sandwich scaffold onto electrodes. Meanwhile, transducer elements, which can be small redox labels, metal ion, or other redox-active species, could report the signal change from target cell binding-induction. In order to achieve high sensitivity and selectivity for cell detection, two mainly signal amplification strategies have been explored. One is based on redox tags such as enzymatic catalyst and metal nanoparticles, and the other is adoption of loading substrate where the graphene and carbon nanotube would be widely employed due to their large surface areas.

Conventional culture plating methods for *E. coli* O157:H7 detection take several days to obtain results, while electrochemical sandwich-type biosensor could provide rapid and sensitive detection [51, 52]. Li et al. have developed a sensitive and efficient electrochemical sandwich assay for detection of *E. coli* (see Fig. 11.7) [51]. Specifically, they immobilized the capture antibodies, which was pre-assembled onto a $SiO_2$-coated AuNPs via a biotin-avidin interaction, onto chitosan-fullerene (C60) composite nanolayer, and then labeled probe antibodies with glucose oxidase (GOD)-loaded Pt nanochains (PtNCs) which served as tracing tag. With such an immunoreaction, they observed a linear detection from 30 to $10^6$ CFU/mL and a LOD of 15 CFU/mL.

Likewise, in another electrochemical immunosensor study, the polypyrrole (PPy)/AuNP/multi-wall carbon nanotube/chitosan hybrid bionanocomposite was employed to modify pencil graphite electrode (PGE) for signal amplification. As such, this platform exhibited a detection linear range from 10 to $10^7$ CFU/mL and detection limit of 30 CFU/mL in PBS buffer [53]. Dos Santos et al. have developed a label-free immunoassay using electrochemical impedance spectroscopy (EIS). They studied the surface antibody functionalization and morphological features by fluorescence and atomic force microscopy. This label-free platform exhibited a detection limit of 2 CFU/mL and a linear range from 30 to $10^4$ CFU/mL [46]. Wang et al. reported a magnetoimmunoassay for rapid separation and sensitive detection of target cells from broth samples [52]. The electrochemical detection of other foodborne pathogens such as *L. monocytogenes*, *Salmonella pullorum*, *S. aureus,* and *Salmonella gallinarum* has been also reported [54–58].

Recently, Zhu et al. have developed an aptamer-cell-aptamer assay for MCF-7 cancer cell detection, employing enzyme label HRP as signal amplification [59].

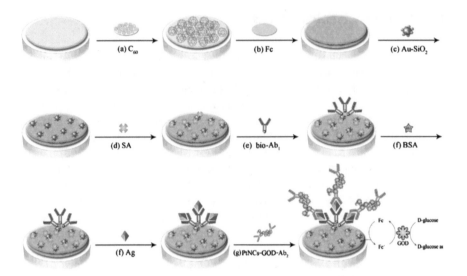

**Fig. 11.7** Schematic description of electrochemical immunoassay for *E. coli* O157:H7 detection. The procedure of the electrode preparation includes five assembling processes, i.e., immobilization of C60, Fc, CHI–SH, Au–SiO$_2$, SA, and bio-Ab1 on the electrode surface. For pathogen detection and signal amplification, the PtNCs-GOD-Ab$_2$ complex was used (Reprinted with the permission from Ref. [51]. Copyright 2013 Elsevier)

Specifically, they fabricated the sensing platform by firstly immobilizing the capture aptamer on Au electrode surface and then capturing target cells followed by an HRP-labeled aptamer. This platform exhibited a detection range from 100 to $1 \times 10^7$ cells/mL, and the detection limit was as low as 100 cells/mL. Likewise, for the detection of the same target cell MCF-7, another study has demonstrated a specific recognition between the aptamer and MUC1 protein that overexpressed on the out surface of the cells [60]. This sensing platform employed aptamer-anchored magnetic beads for cell separations and capture with high selectivity and employed Ag-coated AuNPs as signal amplification. This architecture has achieved a linear detection range between $10^3$ and $10^5$ cells/mL, and the LOD for MCF-7 cell was estimated to be 500 cells/mL. Ge et al. have demonstrated a detection method for the determination of K-562 cells, chronic myelogenous leukemia cells, based on intrinsic peroxidase-like catalytic activity of trimetallic dendritic Au@PdPt nanoparticles, achieving a detection range from $1.0 \times 10^2$ to $2.0 \times 10^7$ cells/mL and a LOD of 31 cells/mL (see Fig. 11.8) [61].

As it is commonly seen for aptamer-based sandwich assay, nanoparticles have been reported for cancer cell detection in electrochemical antibody-based sandwich assay. Chandra et al. developed an electrochemical-sensing platform for drug-resistant cancer cells detection based on Permeability glycoprotein (P-gp) antigen–antibody interaction [62]. Employing Au nanoparticles for loading monoclonal P-gp antibody and hydrazine-labeled carbon nanotube as reduction catalyst, this assay exhibited a linear range from 50 to $1.0 \times 10^5$ cells/mL with the

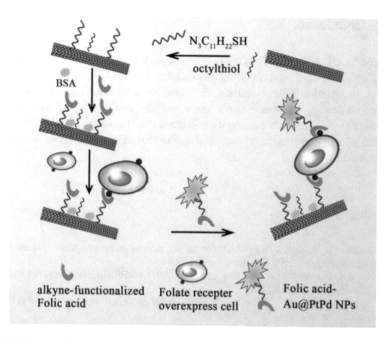

$N_3C_{11}H_{22}SH$

octylthiol

BSA

alkyne-functionalized
Folic acid

Folate recepter
overexpress cell

Folic acid-
Au@PtPd NPs

**Fig. 11.8** Schematic representation of electrochemical sensor of cancer cells by using folic acid functionalized Au@PtPd NPs on paper device. A LOD value of 31 cell/mL has been achieved (Reprinted with the permission from Ref. [61]. Copyright 2013 Elsevier)

detection limit of 2000 cells/mL. In a more recent study, the same research group has further developed a similar platform, again, via employing AuNP as loading support and hydrazine as reduction catalyst, for the determination of metastatic cancer cells. This platform, likewise, achieved a wide linear range between 45 and $1.0 \times 10^5$ cells/mL [63].

## 11.4   Mechanical Biosensors

Sandwich assay-based mechanical biosensors are currently underdeveloped area, in contrast to the optical and electrochemical approaches, for the detection of pathogen and whole cell. Quartz crystal microbalance (QCM), a mechanical technique, relies on a mass variation per unit area by measuring the change in frequency of a quartz crystal resonator. Tothill's group has recently demonstrated a QCM approach based on AuNPs amplified sandwich-type assays for the rapid and real-time detection of bacterial pathogens [64, 65]. For the detection of *Salmonella*, they observed a LOD value at 10 to 20 CFU/mL, while sensing *Campylobacter jejuni*, the sensitivity was 150 CFU/mL.

## 11.5   Conclusion

The applications of sandwich assay biosensors for whole-cell detection are growing rapidly and, as described throughout this chapter, they have been incorporated with different recognition agents and signal transducers. Further improvement of the architecture design, increase bio-receptor selectivity and stability of the assay, and enhancement of transducer sensitivity will pave way for selective, sensitive, rapid, and cost-effective detection of bacterial pathogens and cancer cells at complex sample matrix.

## References

1. Anderson NL, Anderson NG (2002) The human plasma proteome—history, character, and diagnostic prospects. Mol Cell Proteomics 1:845–867
2. Zhang BH, Pan XP, Cobb GP, Anderson TA (2007) microRNAs as oncogenes and tumor suppressors. Dev Biol 302:1–12
3. Ahmed A, Rushworth JV, Hirst NA, Millner PA (2014) Biosensors for whole-cell bacterial detection. Clin Microbiol Rev 27:631–646
4. Li BM, Yu QL, Duan YX (2015) Fluorescent labels in biosensors for pathogen detection. Crit Rev Biotechnol 35:82–93
5. Burlage RS, Tillmann J (2017) Biosensors of bacterial cells. J Microbiol Methods 138:2–11
6. Yetisen AK, Akram MS, Lowe CR (2013) Paper-based microfluidic point-of-care diagnostic devices. Lab Chip 13:2210–2251
7. Kosaka PM, Pini V, Ruz JJ, da Silva RA, Gonzalez MU, Ramos D, Calleja M, Tamayo J (2014) Detection of cancer biomarkers in serum using a hybrid mechanical and optoplasmonic nanosensor. Nat Nanotechnol 9:1047–1053
8. Hsieh K, Ferguson BS, Eisenstein M, Plaxco KW, Soh HT (2015) Integrated electrochemical microsystems for genetic detection of pathogens at the point of care. Acc Chem Res 48:911–920
9. Shen JW, Li YB, Gu HS, Xia F, Zuo XL (2014) Recent development of sandwich assay based on the nanobiotechnologies for proteins, nucleic acids, small molecules, and ions. Chem Rev 114:7631–7677
10. Giri B, Pandey B, Neupane B, Ligler FS (2016) Signal amplification strategies for microfluidic immunoassays. TrAC-Trends Anal Chem 79:326–334
11. Ye DK, Zuo XL, Fan CH (2017) DNA nanostructure-based engineering of the biosensing interface for biomolecular detection. Prog Chem 29:36–46
12. Zhu CZ, Yang GH, Li H, Du D, Lin YH (2015) Electrochemical sensors and biosensors based on nanomaterials and nanostructures. Anal Chem 87:230–249
13. Liang K, Liu F, Fan J, Sun D, Liu C, Lyon CJ, Bernard DW, Li Y, Yokoi K, Katz MH, Koay EJ, Zhao Z, Hu Y (2017) Nanoplasmonic quantification of tumour-derived extracellular vesicles in plasma microsamples for diagnosis and treatment monitoring. Nat Biomed Eng 1:0021
14. Blažková M, Javůrková B, Fukal L, Rauch P (2011) Immunochromatographic strip test for detection of genus *Cronobacter*. Biosens Bioelectron 26:2828–2834
15. Sharma H, Mutharasan R (2013) *hly*A gene-based sensitive detection of *listeria monocytogenes* using a novel cantilever sensor. Anal Chem 85:3222–3228

16. Li FY, Peng J, Zheng Q, Guo X, Tang H, Yao SZ (2015) Carbon nanotube-polyamidoamine dendrimer hybrid-modified electrodes for highly sensitive electrochemical detection of microRNA24. Anal Chem 87:4806–4813
17. Huertas CS, Carrascosa LG, Bonnal S, Valcárcel J, Lechuga LM (2016) Quantitative evaluation of alternatively spliced mRNA isoforms by label-free real-time plasmonic sensing. Biosens Bioelectron 78:118–125
18. Miranda OR, Li X, Garcia-Gonzalez L, Zhu Z-J, Yan B, Bunz UHF, Rotello VM (2011) Colorimetric bacteria sensing using a supramolecular enzyme–nanoparticle biosensor. J Am Chem Soc 133:9650–9653
19. Farrow B, Hong SA, Romero EC, Lai B, Coppock MB, Deyle KM, Finch AS, Stratis-Cullum DN, Agnew HD, Yang S, Heath JR (2013) A chemically synthesized capture agent enables the selective, sensitive, and robust electrochemical detection of anthrax protective antigen. ACS Nano 7:9452–9460
20. Ahmad M, Ameen S, Siddiqi TO, Khan P, Ahmad A (2016) Live cell monitoring of glycine betaine by FRET-based genetically encoded nanosensor. Biosens Bioelectron 86:169–175
21. Zhu LJ, He J, Cao XH, Huang KL, Luo YB, Xu WT (2016) Development of a double-antibody sandwich ELISA for rapid detection of *Bacillus Cereus* in food. Sci Rep 6:16092
22. Vikesland PJ, Wigginton KR (2010) Nanomaterial enabled biosensors for pathogen monitoring—a review. Environ Sci Technol 44:3656–3669
23. Chen ZH, Liu Y, Wang YZ, Zhao X, Li JH (2013) Dynamic evaluation of cell surface N-glycan expression via an electrogenerated chemiluminescence biosensor based on concanavalin a-integrating gold-nanoparticle-modified $Ru(bpy)_3^{2+}$-doped silica nanoprobe. Anal Chem 85:4431–4438
24. Yang HY, Li ZJ, Shan M, Li CC, Qi HL, Gao Q, Wang JY, Zhang CX (2015) Electrogenerated chemiluminescence biosensing for the detection of prostate PC-3 cancer cells incorporating antibody as capture probe and ruthenium complex-labelled wheat germ agglutinin as signal probe. Anal Chim Acta 863:1–8
25. Zhang Y, Tan C, Fei RH, Liu XX, Zhou Y, Chen J, Chen HC, Zhou R, Hu YG (2014) Sensitive chemiluminescence immunoassay for *E. coli* O157:H7 detection with signal dual-amplification using glucose oxidase and laccase. Anal Chem 86:1115–1122
26. Zhang LS, Huang R, Liu WP, Liu HX, Zhou XM, Xing D (2016) Rapid and visual detection of *Listeria* monocytogenes based on nanoparticle cluster catalyzed signal amplification. Biosens Bioelectron 86:1–7
27. Gao LZ, Zhuang J, Nie L, Zhang JB, Zhang Y, Gu N, Wang TH, Feng J, Yang DL, Perrett S, Yan XY (2007) Intrinsic peroxidase-like activity of ferromagnetic nanoparticles. Nat Nanotechnol 2:577–583
28. Wei H, Wang E (2013) Nanomaterials with enzyme-like characteristics (nanozymes): next-generation artificial enzymes. Chem Soc Rev 42:6060–6093
29. Jain S, Chattopadhyay S, Jackeray R, Abid C, Kohli GS, Singh H (2012) Highly sensitive detection of *Salmonella typhi* using surface aminated polycarbonate membrane enhanced-ELISA. Biosens Bioelectron 31:37–43
30. Dykman L, Khlebtsov N (2012) Gold nanoparticles in biomedical applications: recent advances and perspectives. Chem Soc Rev 41:2256–2282
31. Saha K, Agasti SS, Kim C, Li XN, Rotello VM (2012) Gold nanoparticles in chemical and biological sensing. Chem Rev 112:2739–2779
32. Lu WT, Arumugam SR, Senapati D, Singh AK, Arbneshi T, Khan SA, Yu HT, Ray PC (2010) Multifunctional oval-shaped gold-nanoparticle-based selective detection of breast cancer cells using simple colorimetric and highly sensitive two-photon scattering assay. ACS Nano 4:1739–1749
33. Wu WH, Li J, Pan D, Li J, Song SP, Rong MG, Zi Li, Gao JM, Lu JX (2014) Gold nanoparticle-based enzyme-linked antibody-aptamer sandwich assay for detection of *Salmonella Typhimurium*. ACS Appl Mater Interfaces 6:16974–16981

34. de la Rica R, Stevens MM (2012) Plasmonic ELISA for the ultrasensitive detection of disease biomarkers with the naked eye. Nat Nanotechnol 7:821–824
35. Chen R, Huang XL, Xu HY, Xiong YH, Li YB (2015) Plasmonic enzyme-linked immunosorbent assay using nanospherical brushes as a catalase container for colorimetric detection of ultralow concentrations of *Listeria monocytogenes*. ACS Appl Mater Interfaces 7:28632–28639
36. Sung D, Yang S (2014) Facile method for constructing an effective electron transfer mediating layer using ferrocene-containing multifunctional redox copolymer. Electrochim Acta 133:40–48
37. Gehring AG, Brewster JD, He YP, Irwin PL, Paoli GC, Simons T, Tu SI, Uknalis J (2015) Antibody microarray for *E. coli* O157:H7 and shiga toxin in microtiter plates. Sensors 15:30429–30442
38. Song MS, Sekhon SS, Shin WR, Kim HC, Min J, Ahn JY, Kim YH (2017) Detecting and discriminating *Shigella sonnei* using an aptamer-based fluorescent biosensor platform. Molecules 22:825
39. Demirkol DO, Timur S (2016) A sandwich-type assay based on quantum dot/aptamer bioconjugates for analysis of *E. coli* O157:H7 in microtiter plate format. Int J Polym Mater Polym Biomater 65:85–90
40. Hu RR, Yin ZZ, Zeng YB, Zhang J, Liu HQ, Shao Y, Ren SB, Li L (2016) A novel biosensor for *Escherichia coli* O157:H7 based on fluorescein-releasable biolabels. Biosens Bioelectron 78:31–36
41. Dogan U, Kasap E, Cetin D, Suludere Z, Boyaci IH, Turkyilmaz C, Ertas N, Tamer U (2016) Rapid detection of bacteria based on homogenous immunoassay using chitosan modified quantum dots. Sens Actuators B-Chem 233:369–378
42. Kong W, Xiong J, Yue H, Fu Z (2015) Sandwich fluorimetric method for specific detection of *Staphylococcus aureus* based on antibiotic-affinity strategy. Anal Chem 87:9864–9868
43. Yanase Y, Hiragun T, Ishii K, Kawaguchi T, Yanase T, Kawai M, Sakamoto K, Hide M (2014) Surface plasmon resonance for cell-based clinical diagnosis. Sensors 14:4948–4959
44. Liu X, Hu YX, Zheng S, Liu Y, He Z, Luo F (2016) Surface plasmon resonance immunosensor for fast, highly sensitive, and in situ detection of the magnetic nanoparticles-enriched *Salmonella enteritidis*. Sens Actuators B-Chem 230:191–198
45. Eum NS, Yeom SH, Kwon DH, Kim HR, Kang SW (2010) Enhancement of sensitivity using gold nanorods-antibody conjugator for detection of *E. coli* O157:H7. Sens Actuators B-Chem 143:784–788
46. Barreiros dos Santos M, Agusil JP, Prieto-Simón B, Sporer C, Teixeira V, Samitier J (2013) Highly sensitive detection of pathogen *Escherichia coli* O157:H7 by electrochemical impedance spectroscopy. Biosens Bioelectron 45:174–180
47. Charlermroj R, Oplatowska M, Gajanandana O, Himananto O, Grant IR, Karoonuthaisiri N, Elliott CT (2013) Strategies to improve the surface plasmon resonance-based immmunodetection of bacterial cells. Microchim Acta 180:643–650
48. Chen HX, Hou YF, Ye ZH, Wang HY, Koh K, Shen ZM, Shu YQ (2014) Label-free surface plasmon resonance cytosensor for breast cancer cell detection based on nano-conjugation of monodisperse magnetic nanoparticle and folic acid. Sens Actuators B-Chem 201:433–438
49. Mousavi M, Chen HY, Hou HS, Chang CYY, Roffler S, Wei PK, Cheng JY (2015) Label-free detection of rare cell in human blood using gold nano slit surface plasmon resonance. Biosensors 5:98–117
50. Akanda MR, Tamilavan V, Park S, Jo K, Hyun MH, Yang H (2013) Hydroquinone diphosphate as a phosphatase substrate in enzymatic amplification combined with electro-chemical–chemical–chemical redox cycling for the detection of *E. coli* O157:H7. Anal Chem 85:1631–1636
51. Li Y, Fang LC, Cheng P, Deng J, Jiang LL, Huang H, Zheng JS (2013) An electrochemical immunosensor for sensitive detection of *Escherichia coli* O157:H7 using C-60 based biocompatible platform and enzyme functionalized Pt nanochains tracing tag. Biosens Bioelectron 49:485–491

52. Wang Y, Alocilja EC (2015) Gold nanoparticle-labeled biosensor for rapid and sensitive detection of bacterial pathogens. J Biol Eng 9:16

53. Guner A, Cevik E, Senel M, Alpsoy L (2017) An electrochemical immunosensor for sensitive detection of *Escherichia coli* O157:H7 by using chitosan, MWCNT, polypyrrole with gold nanoparticles hybrid sensing platform. Food Chem 229:358–365

54. Cheng CN, Peng Y, Bai JL, Zhang XY, Liu YY, Fan XJ, Ning BA, Gao ZX (2014) Rapid detection of *Listeria monocytogenes* in milk by self-assembled electrochemical immunosensor. Sens Actuators B-Chem 190:900–906

55. Abbaspour A, Norouz-Sarvestani F, Noon A, Soltani N (2015) Aptamer-conjugated silver nanoparticles for electrochemical dual-aptamer-based sandwich detection of *staphylococcus aureus*. Biosens Bioelectron 68:149–155

56. Chen Q, Lin JH, Gan CQ, Wang YH, Wang D, Xiong YH, Lai WH, Li YT, Wang MH (2015) A sensitive impedance biosensor based on immunomagnetic separation and urease catalysis for rapid detection of *Listeria monocytogenes* using an immobilization-free interdigitated array microelectrode. Biosens Bioelectron 74:504–511

57. Fei JF, Dou WC, Zhao GY (2015) A sandwich electrochemical immunosensor for Salmonella pullorum and *Salmonella gallinarum* based on a screen-printed carbon electrode modified with an ionic liquid and electrodeposited gold nanoparticles. Microchim Acta 182:2267–2275

58. Chen Q, Wang D, Cai GZ, Xiong YH, Li YT, Wang MH, Huo HL, Lin JH (2016) Fast and sensitive detection of foodborne pathogen using electrochemical impedance analysis, urease catalysis and microfluidics. Biosens Bioelectron 86:770–776

59. Zhu XL, Yang JH, Liu M, Wu Y, Shen ZM, Li GX (2013) Sensitive detection of human breast cancer cells based on aptamer-cell-aptamer sandwich architecture. Anal Chim Acta 764:59–63

60. Zhang JJ, Cheng FF, Zheng TT, Zhu JJ (2017) Versatile aptasensor for electrochemical quantification of cell surface glycan and naked-eye tracking glycolytic inhibition in living cells. Biosens Bioelectron 89:937–945

61. Ge SG, Zhang Y, Zhang L, Liang LL, Liu HY, Yan M, Huang JD, Yu JH (2015) Ultrasensitive electrochemical cancer cells sensor based on trimetallic dendritic Au@PtPd nanoparticles for signal amplification on lab-on-paper device. Sens Actuators B-Chem 220:665–672

62. Chandra P, Noh HB, Pallela R, Shim YB (2015) Ultrasensitive detection of drug resistant cancer cells in biological matrixes using an amperometric nanobiosensor. Biosens Bioelectron 70:418–425

63. Pallela R, Chandra P, Noh HB, Shim YB (2016) An amperometric nanobiosensor using a biocompatible conjugate for early detection of metastatic cancer cells in biological fluid. Biosens Bioelectron 85:883–890

64. Salam F, Uludag Y, Tothill IE (2013) Real-time and sensitive detection of *Salmonella Typhimurium* using an automated quartz crystal microbalance (QCM) instrument with nanoparticles amplification. Talanta 115:761–767

65. Masdor NA, Altintas Z, Tothill IE (2016) Sensitive detection of *Campylobacter jejuni* using nanoparticles enhanced QCM sensor. Biosens Bioelectron 78:328–336

# Chapter 12
# Biosensors Based on Supersandwich Assays

Xiaojin Zhang and Fan Xia

**Abstract** Only one signal probe is usually bound to the target in traditional sandwich assays, which limits the detection sensitivity. To overcome this limitation, supersandwich assays amplifying the signal through integration of multiple signal probes together have been developed in recent years. In this chapter, we highlight biosensors based on supersandwich assays for the detection of proteins, nucleic acids, small molecules, ions, and cells by a series of efforts reported in the past decade. The detection technologies employed in design of biosensors based on supersandwich assays contain electrochemical assay, electrochemiluminescence assay, fluorescence assay, and surface plasmon resonance assay.

**Keywords** Supersandwich assays · Multiple signal probes · Electrochemical assay
Electrochemiluminescence assay · Fluorescence assay · Surface plasmon resonance assay

The original version of this chapter was revised: Foreword has been included and authors' affiliations have been updated. The erratum to this chapter is available at https://doi.org/10.1007/978-981-10-7835-4_13

X. Zhang (✉) · F. Xia
Engineering Research Center of Nano-Geomaterials of Ministry of Education,
Faculty of Materials Science and Chemistry, China University of Geosciences,
Wuhan 430074, People's Republic of China
e-mail: zhangxj@cug.edu.cn

F. Xia
e-mail: xiafan@cug.edu.cn; xiafan@hust.edu.cn

F. Xia
Hubei Key Laboratory of Bioinorganic Chemistryand Materia Medica,
School of Chemistry and Chemical Engineering, Huazhong University of Science
and Technology, Wuhan 430074, People's Republic of China

## 12.1  Introduction

The sandwich assays have achieved great success in detecting proteins, nucleic acids, small molecules, ions, and cells [1]. Usually, only one signal probe specifically hybridizes with a target in a traditional sandwich assay. Therefore, traditional sandwich assays show relatively low sensitivity because the total signal gain is limited. To overcome this limitation, some approaches combining multiple signal probes together in a sandwich assay to amplify the detection signal have been developed as a kind of sandwich assays, namely, supersandwich assays.

The early and classic example of a supersandwich assay was pioneered by Xia, Zuo, Plaxco, and Heeger in 2010, as shown in Fig. 12.1 [2]. Aiming at the limitation that a target hybridizes with a signal probe in a traditional sandwich assay, they innovatively made a modified signal probe that contained a methylene blue (a redox moiety) label and a "sticky end." The target is hybridized with the signal probe, and the sticky end remained free, which can hybridize with another target.

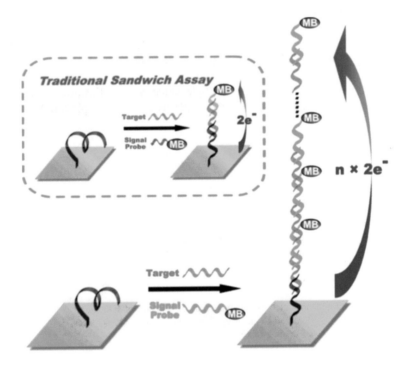

**Fig. 12.1**  Classic example of a supersandwich assay in which the signal probes hybridize to both ends of the target probe to generate long concatamers, which possess multiple target molecules and signal probes. Inset is the scheme of the traditional sandwich assay. (Reprinted with the permission from Ref. [2]. Copyright 2010 American Chemical Society)

Finally, a supersandwich DNA structure with multiple labels was created. This approach led to a significant improvement in detection limit, compared to a traditional sandwich assay. The former had a detection limit of 100 fM, which the latter had a detection limit of 100 pM.

After that, supersandwich assays have been booming in development [3]. In addition to traditional sandwich assays, supersandwich assays have also been widely used in the detection of proteins, nucleic acids, small molecules, ions, and cells. Similarly, electrochemical assay, electrochemiluminescence assay, fluorescence assay, and surface plasmon resonance assay have still been employed in biosensors based on supersandwich assays. According to the detection objects, we divide the assays into four categories: protein detection, nucleic acids detection, small-molecule and ion detection, and cell detection. In each category, the highlighted examples are classified on the basis of the detection technologies.

## 12.2 Supersandwich Assays for Protein Detection

### 12.2.1 Electrochemical Supersandwich Assays

In 2011, Wang et al. proposed an electrochemical immunosensors based on supersandwich multienzyme-DNA label for the detection of Interleukin-6 (IL-6) as a model protein, a biomarker for several types of cancer [4]. The sequence 1 (S1) was conjugated to the secondary antibodies (anti-IL-6) through binding streptavidin of S1 to the biotin tag of anti-IL-6. Then, the capture probe S1 was hybridized with the signal probe S2 with horseradish peroxidase (HRP), which was further hybridized with the target DNA S3, to afford supersandwich multienzyme-DNA label. Supersandwich DNA structure significantly enhanced the amperometric signal, thus achieving a detection limit of 0.05 pg mL$^{-1}$ relative to that of 5.0 pg mL$^{-1}$ using the traditional sandwich label. They then designed an electrochemical biosensor for the detection of folate receptor based on the protecting effect of folate receptor toward folic acid-modified DNA and the signal amplification of supersandwich DNA structure to achieve a detection limit of 0.3 ng mL$^{-1}$, which approached clinically relevant concentrations of folate receptor [5]. They also described an electrochemical biosensor for the detection of thrombin with a detection limit of 10 pM based on G-quadruplex-linked supersandwich structure [6]. Wang et al. used the aptamer with the high affinity to fabricate a label-free supersandwich electrochemical biosensor for the detection of myoglobin, one of the early biomarkers to increase after acute myocardial infarction, based on target-induced aptamer displacement with a detection limit of 10 pM, which was lower than that of those previous antibody-based biosensors for the detection of myoglobin [7].

The high affinity of the negative phosphate backbone of DNA to positively charged metal cations provides an approach to construct metal nanoclusters/nanoparticles along with the DNA template [8]. The metal nanoclusters/nanoparticles possess mimics' enzyme activity, thus having been paid more and

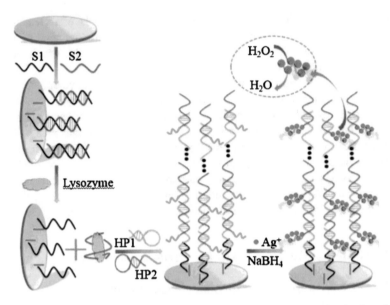

**Fig. 12.2** Schematic illustration of an electrochemical supersandwich assay for the detection of protein lysozyme. DNA S1 and DNA S2 are assembled on the gold electrode. S2 is removed by lysozyme and the left S1 triggered the further HCR process. In the presence of $Ag^+$ and $NaBH_4$, DNA/AgNCs are yielded on the supersandwich DNA structure. Based on the peroxidase-like character of DNA/AgNCs, the lysozyme could be detected. (Reprinted with the permission from Ref. [10]. Copyright 2015 The Royal Society of Chemistry)

more attention in recent years [9]. Wang et al. fabricated an amplified electrochemical aptasensor for the detection of lysozyme based on the mimic oxidase catalytic character of DNA-stabilized silver nanoclusters and hybridization chain reactions (HCR) for signal amplification, as shown in Fig. 12.2 [10]. The DNA duplex was anchored onto the gold electrode and then S2 was specially bound by lysozyme. The left S1 on the surface of the gold electrode triggered HCR of HP1 and HP2 to generate supersandwich DNA structure. $Ag^+$ attached to the cytosine-rich sequence on the 3'-end of HP2 was reduced by $NaBH_4$ to generate DNA/Ag nanoclusters, which had the peroxidase-like character for the detection of lysozyme with a detection limit of 42 pM. Recently, a supersandwich electrochemical immunoassay based on in situ DNA template-synthesized Pd nanoparticles as signal label was proposed through hybridization proximity-regulated catalytic DNA hairpin assembly strategy for the detection of carcinoembryonic antigen with a detection limit of $0.52 \times 10^{-16}$ g mL$^{-1}$ [11].

## 12.2.2 Electrochemiluminescence Supersandwich Assays

DNA methylation plays a significant role in the epigenetic regulation of genomic imprinting, X chromosome inactivation, aging, and carcinogenesis [12]. DNA methylation has become a potential tumor biomarker for a variety of diseases [13]. DNA methylation generally occurs at cytosines in CpG dinucleotides in the mammalian genome along with the catalysis of DNA methyltransferases (MTase) [14]. Therefore, the detection of MTase activity is of significant importance for early cancer diagnosis [15, 16]. Li et al. developed a label-free supersandwich electrochemiluminescence (ECL) assay for the detection of DNA methylation and the methyltransferase activity with a detection limit of $3 \times 10^{-6}$ U mL$^{-1}$ [17]. The cytosine residues of supersandwich DNA structure immobilized on the surface of the gold electrode were methylated through introducing M. SssI and S-adenosylmethionine. Using *Hpa*II endonuclease cleaved the un-methylated cytosines, causing the decrease of ECL signal that was derived from Ru(phen)$_3^{2+}$ (an ECL reagent) intercalated into the grooves of dsDNA. Recently, they reported an ultrasensitive ECL biosensor for the detection of DNA demethylase activity through combining MoS$_2$ nanocomposite with supersandwich DNA structure [18]. A label-free, sensitive, and signal-on ECL assay for the detection of MTase activity with a detection limit of $6.4 \times 10^{-3}$ U mL$^{-1}$ was developed [19]. The methylation of the dsDNA probes on the sensing electrode inactivated the restriction enzyme activity and inhibited subsequent HCR, resulting in the recovery of the ECL signal of the oxygen/persulfate (O$_2$/S$_2$O$_8{}^{2-}$) system.

Ru(phen)$_3^{2+}$ and its derivatives are well-known ECL luminophores that could be intercalated into the grooves of dsDNA [20]. Yuan et al. fabricated a supersandwich ECL assay for the detection of thrombin with a detection limit of 1.6 fM based on Ru(phen)$_3^{2+}$-functionalized hollow gold nanoparticles as signal-amplifying tags [21]. They further employed histidine as a co-reactant of Ru(bpy)$_3^{2+}$ to amplify ECL signal to fabricate a supersandwich ECL assay for the detection of carcinoembry-onic antigen with a detection limit of 33.3 fg mL$^{-1}$ [22]. They also developed a supersandwich ECL assay based on mimic-intramolecular interaction for the detection of prostate-specific antigen (PSA) with a detection limit of 4.2 fg mL$^{-1}$, as shown in Fig. 12.3 [23]. MWCNTs@PDA-AuNPs bound capture antibody (Ab$_1$). The PAMAM dendrimer conjugated Ab$_2$ and supersandwich DNA structure. The detection antibody PSA was immobilized between Ab$_1$ and Ab$_2$. The ECL luminophore Ru(dcbpy)$_3^{2+}$ and co-reactant (histidine) were integrated into super-sandwich DNA structure to amplify the ECL signal.

## 12.3  Supersandwich Assays for Nucleic Acid Detection

### 12.3.1  Electrochemical Supersandwich Assays

The creative case of supersandwich electrochemical assay for nucleic acid detection was reported by Xia et al. [2] as described in the introduction. Inspired by their work, numerous efforts focusing on the electrochemical biosensors for the detection of nucleic acid based on supersandwich assays have been employed so far [24–31]. As an example, an electrochemical biosensor was developed for the detection of

**Fig. 12.3** Schematic
illustration of an ECL
supersandwich assay for the
detection of protein PSA. The
glassy carbon electrode is
modified by
MWCNTs@PDA-AuNPs.
A PAMAM dendrimer is used
to immobilize the detection
antibody and supersandwich
DNA structure. The
supersandwich DNA structure
containing multiple Ru
(dcbpy)$_3^{2+}$ and histidine
further amplifies the ECL
signal. (Reprinted with the
permission from Ref. [23].
Copyright 2014 The Royal
Society of Chemistry)

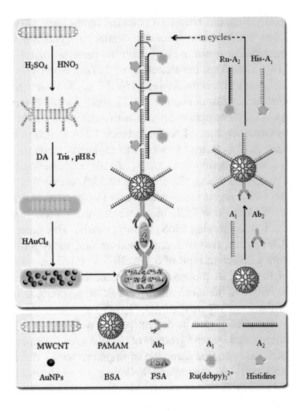

microRNA (miRNA) based on a catalytic hairpin assembly and supersandwich amplification [31]. The target miRNA-221 (a potentially useful biomarker of cancers) triggered the assembly of molecular beacons H1 and H2 to form H1–H2 complexes followed by releasing miRNA-221. H1–H2 complexes were captured on the electrode and further hybridized with HRP-DNAs as signal tags to produce supersandwich DNA structure on the electrode. The reaction of 3,3′, 5,5′-tetramethylbenzidine (TMB)/$H_2O_2$ was catalyzed by HRP to generate amperometric signals that were corresponding to the target miRNA-221. The isothermal dual-amplification strategies without nanoparticles provided high sensitivity and selectivity during detection.

An interesting example that supersandwich DNA structure was constructed on the nanochannel walls to fabricate supersandwich electrochemical assay for the detection of DNA was reported by Xia et al. in 2013, as shown in Fig. 12.4 [32]. The capture DNA probe was first immobilized onto the nanopores and captured the target DNA through hybridization. The signal probes (S1 and S2) were hybridized to create long concatamers, supersandwich DNA structure in the nanopores, which efficiently blocked the pathway for ion conduction. This assay achieved a detection limit of 10 fM for oligonucleotides.

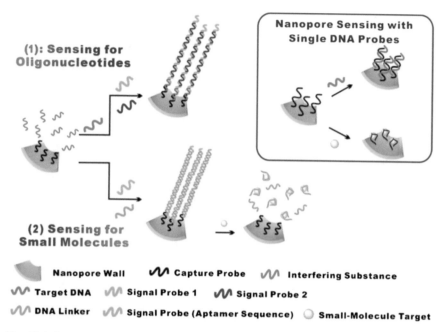

**Fig. 12.4** Schematic illustration of an electrochemical supersandwich assay for the detection of DNA. Inset is the scheme of the traditional sandwich assay in which a single capture DNA hybridizes to a single target strand or binds to a single molecular target. The supersandwich electrochemical assay integrates a more complex DNA nanostructure within the nanopores. (Reprinted with the permission from Ref. [32]. Copyright 2013 Wiley-VCH Verlag GmbH & Co. KGaA, Weinheim)

## 12.3.2   Electrochemiluminescence Supersandwich Assays

Zhang et al. described a highly sensitive supersandwich ECL assay for the detection of the human immunodeficiency virus-1 (HIV-1) gene, as shown in Fig. 12.5 [33]. The capture probe was first anchored on the surface of the gold electrode and hybridized with the target HIV-1 gene. Two auxiliary probes were hybridized with the target HIV-1 gene to generate supersandwich DNA structure on the surface of the electrode. Ru(phen)$_3^{2+}$ as the ECL indicator was intercalated into the grooves of supersandwich DNA structure. The ECL intensity was corresponding to the concentration of the HIV-1 gene with a detection limit of 0.022 pM.

## 12.3.3   Fluorescence Supersandwich Assays

Luminescent silver nanoclusters (AgNC) synthesized using DNA as scaffolds could be acted as fluorescent labels [34]. Wang et al. fabricated a supersandwich DNA/AgNC luminescent sensor through the artificial oligonucleotide scaffold with AgNC biomineralizing unit and target DNA recognizing unit [35]. The recognizing unit

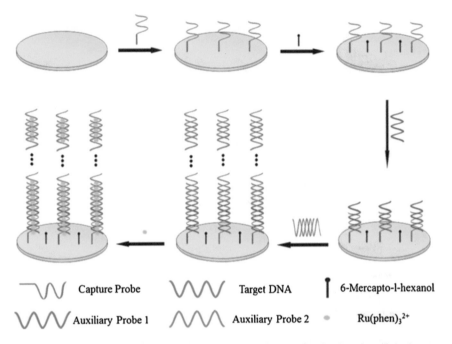

**Fig. 12.5** Schematic illustration of an ECL supersandwich assay for the detection of the human immunodeficiency virus-1 (HIV-1) gene. The high sensitivity and selectivity of electrochemilumi-nescence DNA biosensor can be largely improved by using supersandwich dsDNA along with ECL indicators. (Reprinted with the permission from Ref. [33]. Copyright 2014 Springer-Verlag Wien)

hybridized with the target DNA to create supersandwich DNA structure. The luminescence intensity of AgNC was relative to the concentration of the target DNA. A supersandwich fluorescence in situ hybridization strategy for the detection of mRNA at the single-cell level was reported recently, as shown in Fig. 12.6 [36]. Three kinds of mRNA were tested. Taking TK1 mRNA as an example, a DNA probe entered the fixed cells and hybridized with the target mRNA. Two fluorescent signal probes were hybridized to form long concatamers, thus amplifying the signal of the target mRNA.

## 12.3.4 Surface Plasmon Resonance Supersandwich Assays

Surface plasmon resonance (SPR) is a powerful technology for label-free, real-time, and in situ detection of biomarkers [37]. Surface plasmon resonance biosensor for label-free detection of miRNA based on supersandwich DNA structure and streptavidin signal amplification has been developed by Ding et al. in 2014 [38]. The capture DNA probes immobilized on the gold electrode selectively captured the target miRNA to form supersandwich DNA structure and then hybridized streptavidin through biotin binding for signal amplification, thus leading to the increase of the SPR signal. The assay showed high sensitivity with a detection limit of

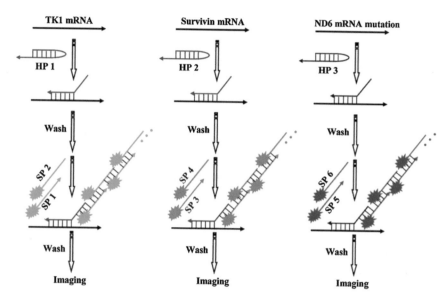

**Fig. 12.6** Schematic illustration of a fluorescence supersandwich assay for the detection of mRNA. Two fluorophore-labeled signal probes are used to generate a supersandwich product, which in turn generates numerous signal probes located at the target mRNA position, resulting in the in situ fluorescence signal amplification. (Reprinted with the permission from Ref. [36]. Copyright 2016 The Royal Society of Chemistry)

miRNA down to 9 pM. Surface plasmon resonance biosensor for enzyme-free detection of miRNA based on supersandwich DNA structure and gold nanoparticles has been proposed by Wang et al. in 2016, as shown in Fig. 12.7 [39]. The capture DNA with a loop immobilized on the gold film surface captured miRNA-21. DNA-linked AuNPs were then captured by hybridization and the report DNAs were hybridized starting from DNA-linked AuNPs to form supersandwich DNA structure, which enhanced the shift of resonance angle. This assay showed high selectivity for the discrimination of single-base mismatch and detected ca. 8 fM miRNA-21. They then lowered a detection limit of miRNA-21 to ca. 0.6 fM by further increase of SPR response using AgNPs absorbed into the grooves of supersandwich DNA structure as additional signal amplification tool [40].

## 12.4    Supersandwich Assays for Small-Molecule and Ion Detection

### 12.4.1    Electrochemical Supersandwich Assays

Adenosine triphosphate (ATP), a small molecule generally acknowledged as a major cellular energy currency, plays an important role in most enzymatic activities

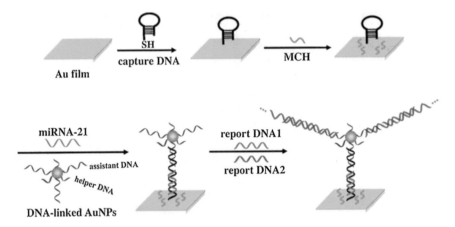

**Fig. 12.7** Schematic illustration of a surface plasmon resonance supersandwich assay for the detection of miRNA. The loop capture DNA immobilized on the Au film surface captures miRNA-21, and DNA-linked AuNPs are then hybridized. The report DNA1 and report DNA2 are introduced to form supersandwich DNA structure. (Reprinted with the permission from Ref. [39]. Copyright 2015 Elsevier)

[41]. ATP depletion is a key process in pathogenesis, particularly, Parkinson's disease, hypoglycemia, and hypoxia [42]. Therefore, the detection of ATP is not only of research interest but of clinical importance [43]. Xia et al. developed an electrochemical aptasensor based on a dual-signaling strategy and a supersandwich assay for the detection of ATP, as shown in Fig. 12.8 [44]. The capture probe anchored on the surface of the gold electrode hybridized with methylene blue (MB)-labeled signal probe and ferrocene (Fc)-modified signal probe to create supersandwich DNA structure. In the presence of ATP, supersandwich DNA structure would disassemble because ATP bounds its aptamer, resulting in the release of the signal probes to generate the reduction signals of MB and Fc. Taking dual signals as the response signal, ATP was detected at a detection limit of 2.1 nM. They also constructed supersandwich DNA structure in the nanopores for the detection of ATP [32, 45]. Of note, other small molecules such as adenosine [46] and cisplatin [47] were also detected by supersandwich electrochemical biosensors with excellent sensitivity and reproducibility.

Mercury(II) ion ($Hg^{2+}$) specifically combines with two thymine bases (T) to afford stable $T\text{-}Hg^{2+}\text{-}T$ bases pairs [48]. Wang et al. fabricated supersandwich DNA assay based on $T\text{-}Hg^{2+}\text{-}T$ to amplify the electrochemical signal for the detection of $Hg^{2+}$ with a detection limit of 10 fM [49]. Silver ion ($Ag^+$), a highly toxic heavy metal ion, has caused serious health and environment attention in recent years [50]. Similar to $T\text{-}Hg^{2+}\text{-}T$, $Ag^+$ specifically combines with two cytosine base (C) to form stable $C\text{-}Ag^+\text{-}C$ bases pairs [51]. Supersandwich electrochemical biosensor based on magnetic nanoparticles labeling with hybridization chain reaction amplification triggered by $C\text{-}Ag^+\text{-}C$ was developed for the detection of $Ag^+$ with a detection limit of 0.5 fM [52]. Recently, the combination of supersandwich DNA structure and

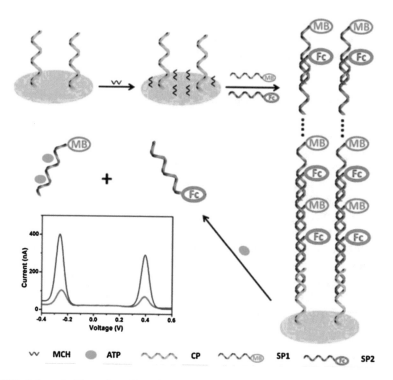

**Fig. 12.8** Schematic illustration of an electrochemical supersandwich assay for the detection of ATP. The capture probe is anchored at the gold electrode surface and then combines with methylene blue (MB)-modified signal probe 1 (SP1) and ferrocene (Fc)-labeled signal probe 2 (SP2) to form supersandwich DNA structure. (Reprinted with the permission from Ref. [44]. Copyright 2016 The Royal Society of Chemistry)

$Zn^{2+}$-requiring DNAzymes in the nanopores provided a strategy to detect $Zn^{2+}$ with a detection limit of 1 nM [53].

## 12.4.2   Electrochemiluminescence Supersandwich Assays

Yuan et al. demonstrated a supersandwich ECL assay for the detection of ochratoxin A (OTA), as shown in Fig. 12.9 [54]. The capture probe immobilized on the surface of the gold electrode triggered a cross-opening process of two hairpin DNAs to form supersandwich DNA structure. Hemin induced the formation of hemin/G-quadruplex DNAzyme structure. In the presence of the target OTA and $RecJ_f$ exonuclease, supersandwich DNA structure disassembled to generate a significant ECL signal of the $O_2/S_2O_8^{2-}$ system. This assay achieved a detection limit of 75 fg $mL^{-1}$ for the detection of OTA. Xu et al. developed a label-free supersandwich ECL assay based on $T$-$Hg^{2+}$-$T$ coordination and the intercalation of Ru (phen)$_3^{2+}$ for the detection of Hg$^{2+}$ [55]. This assay achieved a detection limit of 0.25 nM for the detection of Hg$^{2+}$, meeting the requirement of U.S. Environmental Protection Agency for Hg$^{2+}$ in drinkable water (<10 nM).

### 12.4.3 Fluorescence Supersandwich Assays

ATP was detected using a label-free fluorescence strategy based on the ligation-triggered supersandwich that was reported by Yang et al., as shown in Fig. 12.10 [56]. First, a dsDNA probe was designed as the substrate of ATP-dependent ligation. SYBR Green I (SG I) as the readout signal was intercalated into the grooves of the dsDNA probe. With the addition of ATP, the recognition of T4 DNA ligase caused the dsDNA probe formed supersandwich DNA structure, resulting in the enhancement of the fluorescence signal. This assay showed a high sensitivity with a detection limit of 200 pM for the detection of ATP. For the detection of Hg$^{2+}$, Xu et al. demonstrated a label-free supersandwich fluorescence assay based on the generation of supersandwich DNA structure by T-Hg$^{2+}$-T with a detection limit of 2.5 nM [57]. Genefinder (GF) intercalated into the grooves of dsDNA was employed as the readout fluorescence signal.

## 12.5 Supersandwich Assays for Cell Detection

The detection of cancer cells has become an increasingly important topic for monitoring the progressions of diseases and diagnosing cancers [58]. Zhu et al. developed a supersandwich assay through signal amplification for the detection of

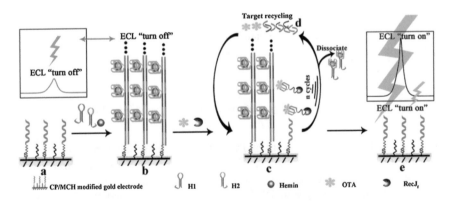

**Fig. 12.9** Schematic illustration of an ECL supersandwich assay for the detection of small molecule OTA. The supersandwich DNA structure is formed on capture probes/6-mercapto-1-hexanol (CP/MCH)-modified gold electrode through a cross-opening process of the two hairpin DNAs. The hemin/G-quadruplex DNAzyme nanostructures are formed upon addition of hemin, the target OTA and RecJ$_f$ exonuclease. (Reprinted with the permission from Ref. [54]. Copyright 2014 The Royal Society of Chemistry)

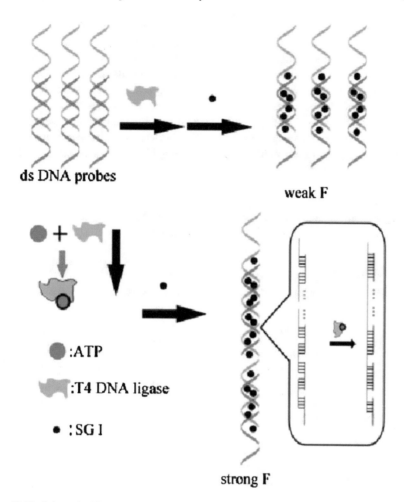

ds DNA probes

weak F

● :ATP

:T4 DNA ligase

● :SG I

strong F

**Fig. 12.10** Schematic illustration of a fluorescence supersandwich assay for the detection of ATP. The double-stranded DNA (dsDNA) probes form a supersandwich which can be detected using dsDNA-specific fluorescent SYBR Green I (SG I). (Reprinted with the permission from Ref. [56]. Copyright 2014 Elsevier)

cancer cells, as shown in Fig. 12.11 [59]. Aptamer-DNA concatamer-quantum dots probes were fabricated by the hybridization of aptamer-DNA and quantum dot-modified DNA with the capture DNA. Multiwall carbon nanotubes (MWCNTs), polydopamine (PDA), and gold nanoparticles (AuNPs) were employed to fabricate the electrode material MWCNTs@PDA@AuNPs through a layer-by-layer method. Concanavalin A (Con A) was captured by multiwall carbon nanotubes@polydopamine@gold nanoparticles (MWCNTs@PDA@AuNPs) that were absorbed on the surface of glassy carbon electrode (GCE). After cancer cells (CCRF-CEM cells) were captured by Con A, aptamer-DNA concatamer-quantum dots probes were modified through the specific recognition of the aptamer to cancer cells. CCRF-CEM cells were detected by both fluorescence and electrochemical methods. The signal amplification of the DNA concatamer and quantum dots improved the sensitivity with a detection limit of 50 cells mL$^{-1}$. In addition, this assay could differentiate cancer cells from normal cells. A supersandwich electrochemical assay based on G-quadruplex DNAzyme and a supersandwich surface plasmon resonance assay using multiple signal amplification strategy were reported for the detection of cancer cells [40, 60]. Furthermore, circulating tumor cells (CTCs), a kind of tumor cells in the peripheral blood, were detected through a

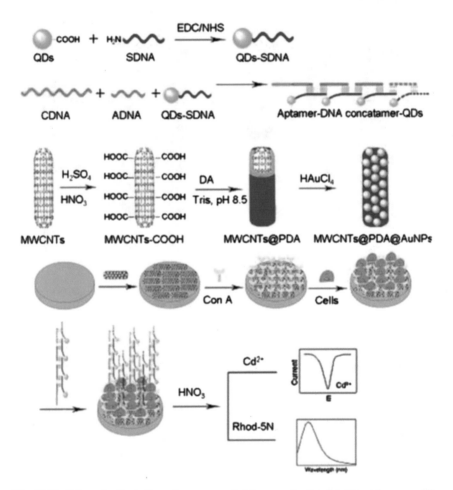

**Fig. 12.11** Schematic illustration of a supersandwich assay for the detection of cancer cells. The DNA concatamer-QDs are designed via DNA hybridization. MWCNTs@PDA@AuNPs composites are assembled to the electrode for immobilization of concanavalin A (Con A). CCRF-CEM cancer cells are selected with aptamer-DNA concatamer-QDs as probes. (Reprinted with the permission from Ref. [59]. Copyright 2013 American Chemical Society)

supersandwich electrochemical assay using signal amplification strategy with a detection limit of 10 cells mL$^{-1}$ [61].

## 12.6 Conclusion

In this chapter, we have summarized the recent development of several biosensors based on supersandwich assays for the detection of proteins, nucleic acids, small molecules, ions, and cells. It is clear that each strategy has its features and limitations. For instance, supersandwich electrochemical assays can detect the target with good sensitivity, but electrochemical detecting instruments are usually required. A facile and sensitive biosensor based on supersandwich assays without complex operations and professional instrumentation will be popularly achieved as immunochromatographic strip. We are happy to see that biosensors based on supersandwich assays step out of the laboratory to the house in the future.

## References

1. Shen JW, Li YB, Gu HS, Xia F, Zuo XL (2014) Recent development of sandwich assay based on the nanobiotechnologies for proteins, nucleic acids, small molecules, and ions. Chem Rev 114:7631–7677
2. Xia F, White RJ, Zuo XL, Patterson A, Xiao Y, Kang D, Gong X, Plaxco KW, Heeger AJ (2010) An electrochemical supersandwich assay for sensitive and selective DNA detection in complex matrices. J Am Chem Soc 132:14346–14348
3. Liu NN, Huang FJ, Lou XD, Xia F (2017) DNA hybridization chain reaction and DNA supersandwich self-assembly for ultrasensitive detection. Sci China-Chem 60:311–318
4. Wang GF, Huang H, Wang BJ, Zhang XJ, Wang L (2012) A supersandwich multienzyme-DNA label based electrochemical immunosensor. Chem Commun 48:720–722
5. Wang GF, He XP, Wang L, Zhang XJ (2013) A folate receptor electrochemical sensor based on terminal protection and supersandwich DNAzyme amplification. Biosens Bioelectron 42:337–341
6. Wang GF, He XP, Zhu YH, Chen L, Wang L, Zhang XJ (2013) G-quadruplex-linked supersandwich DNA structure for electrochemical amplified detection of thrombin. Electroanalysis 25:1960–1966
7. Wang Q, Liu W, Xing YQ, Yang XH, Wang KM, Jiang R, Wang P, Zhao Q (2014) Screening of DNA aptamers against myoglobin using a positive and negative selection units integrated microfluidic chip and its biosensing application. Anal Chem 86:6572–6579
8. Liu HP, Chen Y, He Y, Ribbe AE, Mao CD (2006) Approaching the limit: can one DNA oligonucleotide assemble into large nanostructures? Angew Chem Int Ed 45:1942–1945
9. Song YJ, Qu KG, Zhao C, Ren JS, Qu XG (2010) Graphene oxide: Intrinsic peroxidase catalytic activity and its application to glucose detection. Adv Mater 22:2206–2210
10. Chen L, Sha L, Qiu YW, Wang GF, Jiang H, Zhang XJ (2015) An amplified electrochemical aptasensor based on hybridization chain reactions and catalysis of silver nanoclusters. Nanoscale 7:3300–3308

11. Zhou FY, Yao Y, Luo JJ, Zhang X, Zhang Y, Yin DY, Gao FL, Wang P (2017) Proximity hybridization-regulated catalytic DNA hairpin assembly for electrochemical immunoassay based on in situ DNA template-synthesized Pd nanoparticles. Anal Chim Acta 969:8–17
12. Smith ZD, Meissner A (2013) DNA methylation: roles in mammalian development. Nat Rev Genet 14:204–220
13. Klutstein M, Nejman D, Greenfield R, Cedar H (2016) DNA methylation in cancer and aging. Cancer Res 76:3446–3450
14. Edwards JR, Yarychkivska O, Boulard M, Bestor TH (2017) DNA methylation and DNA methyltransferases. Epigenetics Chromatin 10:23
15. Duan XR, Liu LB, Feng FD, Wang S (2010) Cationic conjugated polymers for optical detection of DNA methylation, lesions, and single nucleotide polymorphisms. Acc Chem Res 43:260–270
16. Flusberg BA, Webster DR, Lee JH, Travers KJ, Olivares EC, Clark TA, Korlach J, Turner SW (2010) Direct detection of DNA methylation during single-molecule, real-time sequencing. Nat Methods 7:461–465
17. Li Y, Luo XE, Yan Z, Zheng JB, Qi HL (2013) A label-free supersandwich electrogenerated chemiluminescence method for the detection of DNA methylation and assay of the methyltransferase activity. Chem Commun 49:3869–3871
18. Sun HP, Ma SX, Li Y, Qi HL (2017) Electrogenerated chemiluminescence biosensing method for the detection of DNA demethylase activity: combining $MoS_2$ nanocomposite with DNA supersandwich. Sens Actuator B-Chem 244:885–890
19. Jiang BY, Yang ML, Yang CY, Xiang Y, Yuan R (2017) Methylation-induced inactivation of restriction enzyme for amplified and signal-on electrochemiluminescence detection of methyltransferase activity. Sens Actuator B-Chem 247:573–579
20. Lei R, Wang XY, Zhu SF, Li N (2011) A novel electrochemiluminescence glucose biosensor based on alcohol-free mesoporous molecular sieve silica modified electrode. Sens Actuator B-Chem 158:124–129
21. Gui GF, Zhuo Y, Chai YQ, Liao N, Zhao M, Han J, Zhu Q, Yuan R, Xiang Y (2013) Supersandwich-type electrochemiluminescenct aptasensor based on $Ru(phen)_3^{2+}$ functionalized hollow gold nanoparticles as signal-amplifying tags. Biosens Bioelectron 47:524–529
22. He Y, Chai YQ, Yuan R, Wang HJ, Bai LJ, Cao YL, Yuan YL (2013) An ultrasensitive electrochemiluminescence immunoassay based on supersandwich DNA structure amplification with histidine as a co-reactant. Biosens Bioelectron 50:294–299
23. He Y, Chai YQ, Yuan R, Wang HJ, Bai LJ, Liao N (2014) A supersandwich electrochemiluminescence immunosensor based on mimic-intramolecular interaction for sensitive detection of proteins. Analyst 139:5209–5214
24. Zhou LY, Zhang XY, Wang GL, Jiao XX, Luo HQ, Li NB (2012) A simple and label-free electrochemical biosensor for DNA detection based on the super-sandwich assay. Analyst 137:5071–5075
25. Chen Y, Wang Q, Xu J, Xiang Y, Yuan R, Chai YQ (2013) A new hybrid signal amplification strategy for ultrasensitive electrochemical detection of DNA based on enzyme-assisted target recycling and DNA supersandwich assemblies. Chem Commun 49:2052–2054
26. Wang J, Shi AQ, Fang X, Han XW, Zhang YZ (2014) Ultrasensitive electrochemical supersandwich DNA biosensor using a glassy carbon electrode modified with gold particle-decorated sheets of graphene oxide. Microchim Acta 181:935–940
27. Wang J, Shi AQ, Fang X, Han XW, Zhang YZ (2015) An ultrasensitive supersandwich electrochemical DNA biosensor based on gold nanoparticles decorated reduced graphene oxide. Anal Biochem 469:71–75
28. Wei BM, Liu NN, Zhang JT, Ou XW, Duan RX, Yang ZK, Lou XD, Xia F (2015) Regulation of DNA self-assembly and DNA hybridization by chiral molecules with corresponding biosensor applications. Anal Chem 87:2058–2062
29. Ren W, Zhou LY, Zhang Y, Li NB, Luo HQ (2016) A reusable and label-free supersandwich biosensor for sensitive DNA detection by immobilizing target-triggered DNA concatamers on ternary self-assembled monolayer. Sens Actuator B-Chem 223:24–29

30. Wei BM, Zhang TC, Ou XW, Li XC, Lou XD, Xia F (2016) Stereochemistry-guided DNA probe for single nucleotide polymorphisms analysis. ACS Appl Mater Interfaces 8:15911–15916

31. Zhang H, Wang Q, Yang XH, Wang KM, Li Q, Li ZP, Gao L, Nie WY, Zheng Y (2017) An isothermal electrochemical biosensor for the sensitive detection of microRNA based on a catalytic hairpin assembly and supersandwich amplification. Analyst 142:389–396

32. Liu NN, Jiang YN, Zhou YH, Xia F, Guo W, Jiang L (2013) Two-way nanopore sensing of sequence-specific oligonucleotides and small-molecule targets in complex matrices using integrated DNA supersandwich structures. Angew Chem Int Ed 52:2007–2011

33. Ruan SP, Li ZJ, Qi HL, Gao Q, Zhang CX (2014) Label-free supersandwich electrogenerated chemiluminescence biosensor for the determination of the HIV gene. Microchim Acta 181:1293–1300

34. Yu JH, Choi S, Dickson RM (2009) Shuttle-based fluorogenic silver-cluster biolabels. Angew Chem Int Ed 48:318–320

35. Wang GF, Zhu YH, Chen L, Wang L, Zhang XJ (2014) Target-induced quenching for highly sensitive detection of nucleic acids based on label-free luminescent supersandwich DNA/silver nanoclusters. Analyst 139:165–169

36. Huang J, Wang H, Yang XH, Yang YJ, Quan K, Ying L, Xie NL, Ou M, Wang KM (2016) A supersandwich fluorescence in situ hybridization strategy for highly sensitive and selective mRNA imaging in tumor cells. Chem Commun 52:370–373

37. Homola J (2008) Surface plasmon resonance sensors for detection of chemical and biological species. Chem Rev 108:462–493

38. Ding XJ, Yan YR, Li SQ, Zhang Y, Cheng W, Cheng Q, Ding SJ (2015) Surface plasmon resonance biosensor for highly sensitive detection of microRNA based on DNA super-sandwich assemblies and streptavidin signal amplification. Anal Chim Acta 874:59–65

39. Wang Q, Liu RJ, Yang XH, Wang KM, Zhu JQ, He LL, Li Q (2016) Surface plasmon resonance biosensor for enzyme-free amplified microRNA detection based on gold nanoparticles and DNA supersandwich. Sens Actuator B-Chem 223:613–620

40. Liu RJ, Wang Q, Li Q, Yang XH, Wang KM, Nie WY (2017) Surface plasmon resonance biosensor for sensitive detection of microRNA and cancer cell using multiple signal amplification strategy. Biosens Bioelectron 87:433–438

41. Patel A, Malinovska L, Saha S, Wang J, Alberti S, Krishnan Y, Hyman AA (2017) ATP as a biological hydrotrope. Science 356:753–756

42. Ma CB, Chen HC, Han R, He HL, Zeng WM (2012) Fluorescence detection of adenosine triphosphate using smart probe. Anal Biochem 429:8–10

43. Huo Y, Qi L, Lv XJ, Lai T, Zhang J, Zhang ZQ (2016) A sensitive aptasensor for colorimetric detection of adenosine triphosphate based on the protective effect of ATP-aptamer complexes on unmodified gold nanoparticles. Biosens Bioelectron 78:315–320

44. Wei BM, Zhang JT, Wang HB, Xia F (2016) A new electrochemical aptasensor based on a dual-signaling strategy and supersandwich assay. Analyst 141:4313–4318

45. Jiang YN, Liu NN, Guo W, Xia F, Jiang L (2012) Highly-efficient gating of solid-state nanochannels by DNA supersandwich structure containing ATP aptamers: a nanofluidic implication logic device. J Am Chem Soc 134:15395–15401

46. Yang XH, Zhu JQ, Wang Q, Wang KM, Yang LJ, Zhu HZ (2012) A label-free and sensitive supersandwich electrochemical biosensor for small molecule detection based on target-induced aptamer displacement. Anal Methods 4:2221–2223

47. Wang GF, He XP, Chen L, Zhu YH, Zhang XJ, Wang L (2013) Conformational switch for cisplatin with hemin/G-quadruplex DNAzyme supersandwich structure. Biosens Bioelectron 50:210–216

48. Ono A, Togashi H (2004) Highly selective oligonucleotide-based sensor for mercury(II) in aqueous solutions. Angew Chem Int Ed 43:4300–4302

49. Wang GF, He XP, Wang BJ, Zhang XJ, Wang L (2012) Electrochemical amplified detection of $Hg^{2+}$ based on the supersandwich DNA structure. Analyst 137:2036–2038

50. AshaRani PV, Mun GLK, Hande MP, Valiyaveettil S (2009) Cytotoxicity and genotoxicity of silver nanoparticles in human cells. ACS Nano 3:279–290

51. Xu G, Wang GF, He XP, Zhu YH, Chen L, Zhang XJ (2013) An ultrasensitive electrochemical method for detection of $Ag^+$ based on cyclic amplification of exonuclease III activity on cytosine-$Ag^+$-cytosine. Analyst 138:6900–6906

52. Zhang YL, Li HY, Chen M, Fang X, Pang PF, Wang HB, Wu Z, Yang WR (2017) Ultrasensitive electrochemical biosensor for silver ion based on magnetic nanoparticles labeling with hybridization chain reaction amplification strategy. Sens Actuator B-Chem 249:431–438

53. Liu NN, Hou RZ, Gao PC, Lou XD, Xia F (2016) Sensitive $Zn^{2+}$ sensor based on biofunctionalized nanopores via combination of DNAzyme and DNA supersandwich structures. Analyst 141:3626–3629

54. Chen Y, Yang ML, Xiang Y, Yuan R, Chai YQ (2014) Binding-induced autonomous disassembly of aptamer-DNAzyme supersandwich nanostructures for sensitive electrochemi-luminescence turn-on detection of ochratoxin A. Nanoscale 6:1099–1104

55. Yuan T, Liu ZY, Hu LZ, Zhang L, Xu GB (2011) Label-free supersandwich electrochemi-luminescence assay for detection of sub-nanomolar $Hg^{2+}$. Chem Commun 47:11951–11953

56. Lin CS, Chen YY, Cai ZX, Zhu Z, Jiang YQ, Yang CJ, Chen X (2015) A label-free fluorescence strategy for sensitive detection of ATP based on the ligation-triggered super-sandwich. Biosens Bioelectron 63:562–565

57. Yuan T, Hu LZ, Liu ZY, Qi WJ, Zhu SY, Aziz ur R, Xu GB (2013) A label-free and signal-on supersandwich fluorescent platform for $Hg^{2+}$ sensing. Anal Chim Acta 793:86–89

58. Galanzha EI, Shashkov EV, Kelly T, Kim JW, Yang LL, Zharov VP (2009) In vivo magnetic enrichment and multiplex photoacoustic detection of circulating tumour cells. Nat Nanotechnol 4:855–860

59. Liu HY, Xu SM, He ZM, Deng AP, Zhu JJ (2013) Supersandwich cytosensor for selective and ultrasensitive detection of cancer cells using aptamer-DNA concatamer-quantum dots probes. Anal Chem 85:3385–3392

60. Lu CY, Xu JJ, Wang ZH, Chen HY (2015) A novel signal-amplified electrochemical aptasensor based on supersandwich G-quadruplex DNAzyme for highly sensitive cancer cell detection. Electrochem Commun 52:49–52

61. Li N, Xiao TY, Zhang ZT, He RX, Wen D, Cao YP, Zhang WY, Chen Y (2015) A 3D graphene oxide microchip and a Au-enwrapped silica nanocomposite-based supersandwich cytosensor toward capture and analysis of circulating tumor cells. Nanoscale 7:16354–16360

# Erratum to: Biosensors Based on Sandwich Assays

Fan Xia, Xiaojin Zhang, Xiaoding Lou and Quan Yuan

## Erratum to:
## F. Xia et al. (eds.), *Biosensors Based on Sandwich Assays*,
## https://doi.org/10.1007/978-981-10-7835-4

In the original version of the book, Foreword has been newly included in Frontmatter and authors' affiliations have been updated throughout. The erratum book has been updated with the changes.

The updated online version of this book can be found at
https://doi.org/10.1007/978-981-10-7835-4_1
https://doi.org/10.1007/978-981-10-7835-4_2
https://doi.org/10.1007/978-981-10-7835-4_3
https://doi.org/10.1007/978-981-10-7835-4_4
https://doi.org/10.1007/978-981-10-7835-4_5
https://doi.org/10.1007/978-981-10-7835-4_7
https://doi.org/10.1007/978-981-10-7835-4_8
https://doi.org/10.1007/978-981-10-7835-4_10
https://doi.org/10.1007/978-981-10-7835-4_11
https://doi.org/10.1007/978-981-10-7835-4_12
https://doi.org/10.1007/978-981-10-7835-4

© Springer Nature Singapore Pte Ltd. 2018
F. Xia et al. (eds.), *Biosensors Based on Sandwich Assays*,
https://doi.org/10.1007/978-981-10-7835-4_13

Printed in the United States
By Bookmasters